Michael Howard was born in 1922 and served with the Coldstream Guards in Italy during the Second World War. He began his academic career in 1947 as Assistant Lecturer in Modern History at King's College London, eventually retiring, as Professor of War Studies, in 1968 to take up a Fellowship at All Souls College, Oxford. In 1977 he was elected Chichele Professor of History of War, and three years later was appointed Regius Professor of Modern History. He left Oxford in 1989 to take up the Robert A. Lovett Chair of Military and Naval History at Yale, from which he retired in 1993. He was a founder of the International Institute for Strategic Studies, of which he remains Life President, and is Vice-President of the Council on Christian Approaches to Defence and Disarmament. He was appointed a Companion of Honour in 2002.

Also by Michael Howard

The Franco-Prussian War

Grand Strategy
(The UK History of the Second World War, vol. IV)

War in European History

War and the Liberal Conscience

The Lessons of History

The First World War

Translation

(with Peter Paret) Clausewitz, *On War*

MICHAEL HOWARD

THE INVENTION OF PEACE AND THE REINVENTION OF WAR

P
PROFILE BOOKS

First published in Great Britain in 2000 by
Profile Books Ltd
3A Exmouth House
Pine Street
Exmouth Market
London EC1R 0JH
www.profilebooks.com

This paperback edition first published in 2001

This revised and extended edition published in 2002

10 9 8 7 6 5 4 3 2

Typeset in Minion by MacGuru Ltd
info@macguru.org.uk

Printed and bound in Great Britain by
Bookmarque Ltd, Croydon, Surrey

A CIP catalogue record for this book is available from the British Library.

ISBN-10 1 86197 409 4
ISBN-13 978 1 86197 409 9

THE INVENTION OF PEACE
AND THE REINVENTION OF WAR

'War appears to be as old as mankind,
but peace is a modern invention.'

Sir Henry Maine (1822–88)
International Law *(London, 1888) p. 8*

CONTENTS

PREFACE

This essay was originally written at the request of Professor David Cannadine, the Director of the Institute for Historical Research of the University of London. It is an extended version of the plenary lecture that inaugurated the Anglo-American Conference on War and Peace held at the Institute in July 2000.

Fourteen months after its delivery there occurred the terrorist attacks in New York and Washington, DC, in which some 3,000 people were massacred and, in the eyes of many, a new and grimmer era opened in the history of mankind. I have added an epilogue reflecting on the significance of that event. In dealing with such *actualités* it is impossible, and perhaps undesirable, to maintain the dispassionate stance for which a historian must strive in analyzing the events of the past, and some readers may find my remarks controversial. For this I make no apology. If anything that I write can make an impact, however minimal, on the course of events, so much the better.

Michael Howard

June 2002

I

INTRODUCTION

'War appears to be as old as mankind, but peace is a modern invention.' So wrote the jurist Sir Henry Maine in the middle of the nineteenth century. There is little to suggest that he was wrong. Archaeological, anthropological, as well as all surviving documentary evidence indicates that war, armed conflict between organized political groups, has been the universal norm in human history. It is hardly necessary to explore whether this was the result of innate aggression, or whether aggressiveness arose from the necessity of fighting for such scarce resources as water or land. Rousseau may have been right in suggesting that men in a mythical state of nature were timid, and only became warlike when they entered into social relations; but social relations were necessary for survival. What Kant termed man's 'asocial sociability' automatically created conflict as well as co-operation.

Peace may or may not be 'a modern invention' but it is

certainly a far more complex affair than war. Hobbes bleakly defined it as a period when war was neither imminent nor actually being fought, but this definition is hardly comprehensive. At best this is what is usually described as negative peace. Often it is the best that people can get, and they are duly thankful for it. But peace as generally understood today involves much more than this. Positive peace implies a social and political ordering of society that is generally accepted as just. The creation of such an order may take generations to achieve, and social dynamics may then destroy it within a few decades. Paradoxically, war may be an intrinsic part of that order, as we shall see. Indeed throughout most of human history it has been accepted as such. The peace invented by the thinkers of the Enlightenment, an international order in which war plays no part, had been a common enough aspiration for visionaries throughout history, but it has been regarded by political leaders as a practicable or indeed desirable goal only during the past two hundred years.

Some societies have certainly been more warlike than others, probably from necessity. In some, war may have originated as religious ritual, or as a rite of passage for adolescents, or as a form of play, like football matches, for adult males in which death was risked but not necessarily inflicted; but ultimately it became a more serious matter for the reason pithily stated by Clausewitz, that if one combatant is prepared to use extreme measures his antagonist has to follow suit. When fighting is necessary for physical survival those who are good at it will predominate. If they pass on their genes to their offspring they will found ruling dynasties. They and their companions become warrior elites whose interests and attitudes determine the nature of their culture, including religion, literature and the arts. They create a social and political order, which initially may have no justification but its own strength, but for which utility, prescription and, above all, religious sanction ultimately provide legitimacy. Legitimized order produces domestic peace, and also legitimizes the conduct of war. Success in war further reinforces legitimacy. Failure results either in subjection and the imposition of an exogenous elite whose rule in turn becomes legitimized by prescription, or the eventual emergence of another indigenous elite more successful than its predecessors.

The greater the effectiveness of a military elite, the greater will be its capacity for extending its power and creating hegemonies. Warriors may go off on their own, as did the Nor(se)mans in the tenth and the Spaniards in the fifteenth century, and establish an imperial hegemony over alien populations. The viability of their rule will initially depend on their continuing military power and will to use it – a will probably, though not invariably, based on a sense of moral superiority derived from religion, race and general culture. But, ultimately, if their dominance is to survive, it must be legitimized: by their success in converting their subjects to their own system of beliefs, by the co-operation of indigenous elites, and above all by their ability to maintain economic and political stability in the societies they govern.

This last is the most important condition of all, and perhaps explains the longevity of such hegemonies as the Ottoman empire and the successive dynasties in China. Change is the greatest enemy of stability and so, in consequence, of peace. In rural societies which change little over centuries if not millennia, prescription ultimately makes any rule acceptable. The main variable lies in the harvests. Bad harvests make it impossible to pay otherwise acceptable taxes and create peasant unrest; but other things being equal, this is usually isolated and suppressible. If other things are not equal, such suppression triggers wider disorders, as it did in Germany in the sixteenth, or the Balkans in the nineteenth, century. This in itself indicated that, for whatever reason, society was no longer stable, and that order could be preserved or restored ultimately only by adjustment to new conditions.

War, it has rightly been said, starts in the minds of men, but so does peace. For some people – perhaps for most – any order is acceptable so long as their expectations are met, and for most of human history these expectations have been very basic. This majority will be little concerned about injustice to others, if indeed they ever hear about it. For them peace is what they have got, and they want to preserve it. There will always be a minority, however small, aware of the imperfections of their societies as measured by standards of divine or natural justice, but such awareness usually demands an exceptional degree of education, leisure and independence. In warrior societies such people were normally either born or co-opted into a priesthood which, whatever absolute standards of behaviour it might advocate, was nonetheless dedicated to the legitimization of the existing order. When the increasing complexity of such societies resulted in a class of educated laity, it was from among their ranks that critics of the social order naturally emerged. Francis Bacon noted at the end of the sixteenth century that one of the causes of sedition in a state was 'breeding more scholars than preferment can take off'. For such critics the oppressions and shortcomings of the existing order rendered it so unjust and illegitimate that both internal rebellion and external war against it was justified. For them, peace could come about only through the creation of a new order. Throughout human history mankind has been divided between those who believe that peace must be preserved, and those who believe that it must be attained.

As we shall see, the medieval order, as it developed in Europe between the eighth and the eighteenth centuries, was largely a matter of a successful symbiosis between the ruling warrior class that provided order and the clerisy that legitimized it. Eventually critics emerged from within that clerisy who denied the essential legitimacy of their rulers on the grounds that war was not a necessary part of the natural or divine order, but a derogation of it. It was then that peace, the visualization of a social order from which war had been abolished, could be said to have been invented; an order, that is, resulting not from some millennial divine intervention that would persuade the lion to lie down with the lamb, but from the forethought of rational human beings who had taken matters into their own hands. The significance of that invention, and the difficulties mankind has found in implementing it, provide the subject matter for the following essay.

II

PRIESTS AND PRINCES: 800–1789

The history of Europe is certainly not typical of world history, and I shall be quite properly taken to task for focusing on it. I do so for two reasons. First, European history is the only field in which I can claim any kind of expertise, and any comments I had to make on other regions would be pretentious, superficial and probably wrong. My second and more serious excuse is that it was in Europe, and its overflow in North America, that there developed the thinking about war and peace that now constitutes the bulk of global discourse about the topic. We still think about peace and how to establish it in terms originating in the European Enlightenment and often little changed since then; about war in categories developed by Clausewitz and Western practice over the past five hundred years; and about the relation between the two in terms to be found in the teaching of Christian churches over two thousand years and Western lawyers over the past four hundred. Any contribution from neither classical nor European sources has been so deeply internalized that I am afraid very few of us are aware that it even exists.

European society during the millennium between the eighth and eighteenth centuries was probably exceptionally bellicose, and that bellicosity does much to explain its eventual global dominance. But it had to be bellicose if it was to survive at all. After the end of the Roman hegemonial peace in the fifth century there came the conflicts consequent on the *Völkerwanderung*, as successive tribes from the east invaded western Europe, settling and displacing existing populations. Over centuries of fighting, warrior leaders emerged who provided local protection and whose families became the nuclei of a society whose structure was predicated on the assumption of permanent war. Their power was stabilized and legitimized, not only by the ever-present threat posed by invading Moslems, Magyars or Vikings, but by the co-operation of a church that provided both a divine sanction for the existing order, and an educated class to provide the sinews of administration.

The church had to solve the problem of reconciling a

doctrine of a divine order, in which all differences were reconciled and to which the concept of peace was basic, with the reality of a war-torn world in which its very survival depended on the protection and favour of successful warlords. The solution had been found by St Augustine in the fourth century. War, he taught, had to be accepted as part of the fallen condition of man, who was simultaneously a citizen of the City of God and of a worldly kingdom which, with all its imperfections, played an essential part in the divine purpose and could therefore rightly impose its own obligations. War against the enemies of Christendom itself was entirely justifiable – the Old Testament provided plentiful justification and guidance as to how to wage it – and even intramural war within Christendom had to be accepted as part of mankind's fallen condition. The latter, however, was intrinsically sinful, and clear limitations were imposed on its conduct. These limitations were refined down the centuries. War had to be waged under a proper authority and as a last resort; to right a wrong; and do no more damage than was essential to the achievement of its purpose. Basically, war had the function of upholding or restoring the secular order sanctified by the church; an order that provided peace, justice and protection for all Christians. Those who fought were serving God's purpose no less than those who prayed and those who worked – a threefold hierarchy of estates which persisted until the leaders of the French revolution replaced it with the concept of a homogenous nation represented only by the third estate and from which the other two were by implication excluded.

War was thus recognized as an intrinsic part of the social and political order, and the warrior was accepted as a servant of God, his sword as a symbol of the Cross. A culture of chivalry developed around the role and activities of the knight, that had little to do with the brute realities of war, and nothing whatsoever with wars against the infidel which could be, and were, fought with unrestrained brutality. This assimilation between warrior and priest was underpinned by the concordat between the most powerful family in western Europe, the Carolingian dynasty, and the surviving Christian church in the West, which was sealed by the coronation in AD 800 of Charlemagne as Holy Roman Emperor. Legitimized both as the heir of the vanished but still respected hegemony of Rome and as the instrument of the church, Charlemagne did not have the power to sustain this notional hegemony beyond his own generation and it was to be repeatedly devolved and divided. Nonetheless, the concept of the Holy Roman

Empire remained one of enormous importance until the Westphalian settlement of 1648, if not until its ultimate demise in 1803. It was the first of the many new world orders, whose somewhat melancholy succession is recorded in the following pages.

*

It is important to understand why the hegemony of the Holy Roman Empire remained from the beginning very largely notional. Mobile forces were necessary to defend its widespread frontiers and enforce authority within them, but the mounted men-at-arms who provided those forces were very expensive to raise, train and maintain. In an economy still deprived of specie, land was the only way of paying for their upkeep, and once land was alienated it was very difficult to get it back again. Whatever oaths of allegiance might be sworn, the land bestowed on tenants-in-chief became a basis for their independent power which could be protected by castles, the reduction of which demanded expensive forces of specialists and prolonged campaigns. So as the tides of the eighth- and ninth-century invaders – Moslem, Magyar, Viking – ebbed or were absorbed, they left a Europe parcelled out between thousands of lords, each with his own power base, owing allegiance to a notional overlord whose authority was effective only so far as he could enforce it. Once there was no external threat to unite them they were free to engage in endless disputes over property rights. War was no longer a struggle for survival but a form of litigation, limited, like all litigation, by the resources of the litigants. In these conflicts there were more losers than winners. A dozen or so families rose to dominance, but the rest of the warrior caste gradually became impoverished and *désoeuvré*, except in so far as the crusades were able to distract them and provide job satisfaction. By the fourteenth century this militant aristocracy had become a source of disorder rather than order, but its members remained socially and culturally dominant long after they had lost political power. Peace they regarded as a brief interval between wars, to be filled with such warlike activities as tournaments, jousting and, increasingly, hunting, to keep them fit for the next serious conflict: a habit that has survived into our own day in the upper-class obsession with hunting and field sports. Nevertheless, this culture had its positive side: a society that disdained such qualities as nobility, honour, loyalty, and indeed chivalry, however much these may have been abused by those who claimed to possess them, would be a sadly impoverished one.

It was very largely because war, for the ruling classes, remained an almost automatic activity, part of the natural order of things, that wars for territory, inheritance and allegiance, sharpened by religious differences, continued well into the seventeenth century. But the longer these wars continued, the more the power of the old nobility was weakened vis-à-vis that of the princely dynasties whose independent power was legitimized by the ritual of kingship and who could thereby call on the direct allegiance of all the inhabitants of their realms and exercise powers of taxation over them to pay for their wars. These wars were becoming increasingly expensive. For anything more than the briefest of campaigns those who conducted them had to enlist paid professionals. The old mounted men-at-arms lost their primacy on the battlefield to foot soldiers trained in the mastery of bows, pikes and, later, muskets. Most significant of all was the development of guns, *ultima ratio regum*, the final argument of kings against over-mighty subjects whose castles could now be reduced to rubble in weeks if not days. But to pay for all this, princes had to raise money, either by loans or by direct taxation of their subjects. The representative bodies of those subjects attempted more or less successfully to exact political concessions in return – usually a confirmation or extension of their rights and privileges. Some of those bodies, as in England, were highly successful in doing so. Elsewhere, especially in Germany, they became historical curiosities. But everywhere, as a result of this growing intercourse between princes and their subjects, there came into being a new entity as the basis for the international order: the state.

*

An essential element in the birth of the state was the disintegration of ecclesiastical support for the medieval structure of power that took place with the Reformation in the early sixteenth century. Dissident clergy provided support for dissident princes, and vice versa. With the co-operation of a church whose spiritual authority extended to the most remote of rural settlements, princes could declare themselves absolute from the higher authority of pope and emperor and command popular support; which was forthcoming, not so much as a result of any growth of national self-consciousness – though this certainly existed in Shakespeare's England – as from simple xenophobia: dislike of a church dominated by foreigners. At the same time, the princes sponsored the growth of an educated laity to staff their increasingly complex bureaucracy, manage a now entirely secular legal system, and in general to exercise and legitimize the power of the state. And with the state, there came into being a new 'new order'; one which political scientists have termed a system of states.

The birth of this new order was exceptionally painful. If European culture in the sixteenth century was becoming secularized, it nonetheless remained bellicose. Indeed, the entire apparatus of the state primarily came into being to enable princes to wage war. With few exceptions, these princes still saw themselves, and were seen by their subjects, essentially as warrior leaders, and they took every opportunity to extend their power. The attempts by the dominant dynasties of Europe to exercise disputed rights of inheritance throughout the fourteenth and fifteenth centuries became consolidated, in the sixteenth century, into a bid by the Habsburgs to sustain a hegemony which they had inherited over most of western Europe against all their foreign rivals and dissident subjects, usually under the leadership of France. The result was almost continuous warfare in western Europe from the early sixteenth until the mid-seventeenth centuries. The expense of this warfare crippled even the most powerful of monarchs. The taxes they imposed to conduct it led to repeated internal revolts. Their unpaid armies mutinied. With bureaucracies inadequate to manage their armed forces they had to devolve their recruitment, pay and leadership on to mercenaries who made war much as they wished and, in default of regular payment, recouped their losses at the expense of the unfortunate peoples among whom they found themselves fighting. The old order had irretrievably broken down, and a new seemed powerless to be born.

The date of birth of this second new world order is generally taken to be 1648, when the Peace of Westphalia effectively affirmed the state as the unchallenged guarantor of domestic order and legitimiser of external war. The history of Europe was henceforward to be shaped by the relations between its states, and the international order depended on their ability to create among themselves an effective international society. In that society war continued to be endemic for reasons we shall consider in a moment, but not the least important was the structure of what was retrospectively to be labelled the *ancien régime*: a mutually supportive trinity of monarchy, church and aristocracy.

The absolute power of the sovereign prince, absolute, that is, from external control, was recognized by the Westphalia settlement, but its internal effectiveness depended on his success in overcoming the attempts of the aristocracy to maintain their own rights, and those of the emerging third estate – consisting largely of merchants, craftsmen and an urban bourgeoisie – to assert new ones. In Prussia the Hohenzollerns were notoriously successful. In England and Scotland the Stuarts, equally notoriously, failed. The Bourbons in France and the Habsburgs in Spain and central Europe, behind a façade of courtly splendour, established uneasy compromises. The authority and power of these sovereigns was competitively displayed in their patronage of the arts and the splendour of their courts. At their zenith they created a superb high culture, the fruits of which in music, architecture and the arts in general we can probably enjoy today a great deal more than did that tiny, overdressed and protocol-ridden minority privileged at the time to share it. Nevertheless, until the mid-eighteenth century these princes still derived their prestige, however remotely, from their status as warrior leaders, and were depicted as such in the court iconography. Nor was this anachronistic. The power, prestige and sometimes the very existence of the state still depended on their military success. It was the military leadership of their rulers that won for Russia and Prussia the rank of European powers, and the failure of such leadership that doomed Sweden to second-class status and Poland to oblivion.

Essential to the power of the monarch was a supportive, if not a submissive, church. The Reformation had created throughout northern Europe Protestant churches which owed their very existence to the princes who they in their turn loyally supported. In southern Europe the Catholic church and the Habsburg dynasty recognized their mutual interdependence, and between them created the culture of the baroque, the awe-inspiring magnificence of whose buildings bludgeoned the imagination of their subjects into a devotion as much to Caesar as to Christ. In France the church put as much distance as possible between itself and the papacy in its willing subordination to the temporal power of the Crown. In return, the monarchs imposed a monopoly of religious dogma that reinforced their own social control over a population whose Christian beliefs had not yet been shaken by the stirrings of free thinking among a tiny group of scholars.

Finally, there was an aristocracy whose local power in what was still an overwhelmingly rural society was very largely absolute, and which the monarch was happy to support so long as they in turn supported him and recognized his unique authority to raise taxes and make war: two functions without which no state could survive, as was shown by the unhappy example of Poland. The magnificence of the royal court was not simply decorative, but functional: it had to outshine, not only that of rival monarchs, but that of potential domestic competition as well. The splendour of the aristocrats had to be seen as a mere reflection of that of the sovereign prince. In return the aristocracy enjoyed a monopoly of court positions, which were multiplied to absurd lengths to accommodate them, and, naturally, of commissions in the royal army; an army now raised, paid, clothed and trained by the state, and swearing allegiance only to the Crown. But there was a heavy price to be paid for all this. The court culture insulated all but the strongest-minded of monarchs from the rest of society at a time when enormous social changes were beginning to take place; while an aristocratic officer class was by instinct hostile to any military developments disruptive of traditional ways of waging war.

*

Such was the pattern of the *ancien régime* as it emerged at the end of the seventeenth century; initially in the France of Louis XIV, but which came to be copied throughout continental Europe. But in the now united kingdoms of England and Scotland, usually inaccurately known as England, a very different system was developing; a system that has persisted until our own day and seems likely to last for some time yet.

In the England, if not the Scotland, of the seventeenth century, the original warrior aristocracy had largely destroyed itself in internecine wars, and been replaced by the Tudors with loyal courtiers and bureaucrats who proved able landowners and administrators but who had not inherited, with their titles and estates, any great inclination or aptitude for war. Fortunately, thanks to Britain's insular position they were not called upon to display any; and where such an inclination did exist, largely among the lesser gentry and squirearchy, it found profitable employment on or beyond the seas. Secure from any serious danger of invasion, the English did not need an army, except to control the rebellious Irish. When a danger of invasion did arise, in 1640, it was only from Scotland, as part of England's own internal religious conflicts. The attempt by the Stuarts to raise an army to deal with the threat from Scotland precipitated a civil war which they comprehensively lost. By the end of the century they had been replaced by an oligarchy of landowners and merchants which subordinated both Crown and church to the interests of what was essentially a bourgeois culture; one concerned with the maximization of wealth rather than military power. An army was needed, partly to control a continually rebellious Ireland, partly to play a part in the power struggles on the continent into which Britain was drawn by Louis XIV's quest for domination, and partly to garrison and defend Britain's growing overseas possessions; while the need for a navy was self-evident. But by a very British compromise both army and navy, although raised and maintained in the name of the Crown and owing allegiance to the person of the monarch, were paid for by Parliament year by year. The armed forces were thus effectively neutralized as instruments of royal power.

This did not mean that the British ruling classes in the eighteenth century were in the least pacific. They may have been bourgeois but they remained bellicose, if only because war increased their wealth. A later generation might associate prosperity with peace, but in the eighteenth century this connection was by no means clear. For the merchants, the lucrative trade with the East and West Indies depended on the fighting efficiency of the Royal Navy. For the naval officers themselves, naval prize money could purchase a substantial country estate. For the landed gentry, a commission in the army provided social prestige, though they had to pay for it. For impoverished Scots of all classes, the army was a royal route to upward mobility. But the officers of the armed forces remained firmly rooted in the oligarchy from which they were drawn, and although their loyalty might in principle be to the throne, they felt little affection for the rather dull Germans who occupied it. They fought formally for their king, but viscerally for their 'country'; a concept embodied in the pugnacious and xenophobe figure of the small farmer John Bull.

*

So the institution of war persisted as part of the international order in eighteenth-century Europe; partly because there were still serious issues of power to be determined, partly because it came naturally both to the ruling classes and to the sovereigns themselves. In eastern Europe there was still serious fighting to be done against the encroaching Turks, while conflict continued on and beyond the seas over access to trade and wealth. But the conduct of war by land was now subjected to stringent economic constraints. Armies now consisted of full-time and highly trained professionals directly paid by the state, and too costly to be wasted on pointless campaigns. Since these armies had not only to be paid but centrally supplied if they were to fight effectively, they required considerable logistical support; which was made all the more necessary by the increasing use of firepower, as artillery became more mobile and effective. The huge supply trains that all this involved made movement difficult along unsurfaced roads that rapidly turned to mud in wet weather, and restricted campaigning to a few months of the year. Further, improvements in the art of fortification made it possible to block major roads and often reduced campaigns to prolonged siege warfare. War was thus prolonged, inconclusive and very expensive. Both France and Prussia were bankrupted by the Seven Years War, after which there was to be no serious conflict in Europe for a quarter of a century. Gibbon happily termed warfare during this period as 'temperate and indecisive', but for the Prussian soldier Clausewitz it was a frustrating time in which armies simply 'consumed themselves', and the French military writer Guibert lamented the decadence of a society apparently no longer capable of serious fighting and at the mercy of anyone who was.

This kind of limited warfare – decadent to some, civilized to others – both made possible and reinforced the Westphalian concept of international order. Before Westphalia, there had still survived something of the medieval concept of authority as being devolved from God through Dantesque gradations in which secular, ecclesiastical and divine hierarchies paralleled and reflected one another. In the second half of the seventeenth century, with the emergence of states whose rulers were both absolute from superior control and immediate to their subjects, a new Newtonian concept came into being, in which order was preserved by the relationship between states themselves, as the order of the universe was preserved by the relationship between the planets. By the beginning of the eighteenth century the preservation of peace came to be seen as the preservation of a balance between powers; a balance that might have to be constantly adjusted by wars. Indeed, an explicit rationale given to Parliament for the maintenance of a standing army in Britain, until the middle of the nineteenth century, was 'the maintenance of the balance of power'. Wars were still seen as a part of the international order, but their conduct had to take account of the balance they were fought to preserve and the settlement with which they were intended to end. The aristocracy might still see war as a normal social activity that needed no justification, but for such 'enlightened despots' as Frederick II of Prussia and Joseph II of Austria, it was simply a necessary tool in the exercise and preservation of state power; though one now so expensive as to be better avoided. States had an intrinsic right to go to war when they thought it necessary, and state policy was a perfectly adequate *jus ad bellum*. The great Dutch international lawyer Hugo Grotius, writing during the first half of the seventeenth century, had tried to secularize the doctrine of the just war developed by the Catholic church, defining the right to go to war in traditional terms of preserving justice and righting wrongs; but his eighteenth-century successor Emer de Vattel taught to willing ears that such reasoning was irrelevant when both parties believed themselves to be in the right and no superior authority existed to adjudicate between them. What mattered was *jus in bello*; to conduct war in such a manner as to do the least possible damage to international society as a whole, and make possible the conclusion of a stable peace.

*

This rationalization of war can also be attributed, at least in part, to social change. As western Europe became wealthier and more urbanized, the bourgeoisie more influential and better educated, and the hold the church exercised over the minds of men gradually diminished, there emerged the group of thinkers who spearheaded that greatest of intellectual revolutions in the history of mankind, the Enlightenment: the rejection of traditional authority in church and state and the elevation of reason and individual judgement as the only acceptable basis for both intellectual and political authority. The leaders of the Enlightenment – the *philosophes* in France, the *Publizisten* in Germany, the Scottish academics, the English Nonconformists – were not only independent of the ruling trinity of the *ancien régime* – monarch, church and aristocracy – but subversive of its entire culture. They saw war not as part of the natural order or a necessary instrument of state power, but as a foolish anachronism, perpetuated only by those who enjoyed or profited by it. Their pacifist sentiment was fuelled not so much by humanitarian revulsion and war-weariness – few of them had any direct experience of war – as by the perception that war was part of an entire social system from which they wished to dissociate themselves and hoped ultimately to destroy.

Paradoxically, it was the very limitation of warfare in the eighteenth century that made this possible. In France, the middle classes were excluded from military service in order to provide careers for an increasingly indigent aristocracy (although it must be noted that members of the lesser nobility and clergy excluded from court favour provided some of the Enlightenment's most notable leaders). In Prussia, they were also exempt, so as to provide a tax-base for the royal wars. This very absence of involvement made it possible for the bourgeoisie to regard war simply as an activity waged by monarchs, aristocrats and the sweepings of society for their own enjoyment and advantage; one not only remote from civilized people, but entirely unnecessary, and one that would inevitably disappear as the rule of reason extended its sway.

In Britain, as we have seen, the situation was different. There the merchant classes profited by war and naturally supported it. But there was also a substantial opposition consisting of the Nonconformists who had been excluded from the governing classes ever since the Restoration of 1660, but who became increasingly influential during the eighteenth century as rapid commercial and industrial development increased their wealth and social acceptability. Standing as they did outside the state machine and excluded from the elites that controlled it, they had no sympathy for an order underwritten by the Anglican church and based on the power of the monarchy and the landed gentry. Inspired partly by enlightened rationalism, but more strongly by a belief in a divine order based not on ecclesiastical authority but on individual conscience and ecumenical brotherhood, they provided the Enlightenment with a powerful religious sanction that it lacked on the continent.

A spore of such people had already settled in North America and there established a society governed by an oligarchy that, uniquely, did not justify its ascendancy by its past or present military activities. Thomas Jefferson's description of it as an 'aristocracy of talent and virtue' may have been pitching it a bit high, but the claim to ascendancy of the highly educated lawyers and landowners who dominated society in the American colonies certainly did not, unlike contemporary European elites, depend on the military prowess of their ancestors. The ruling philosophy of the generation that established the independence of the United States was the very quintessence of the Enlightenment, with its belief in the rights and perfectibility of man and his capacity for peaceful self-government once the artificial barriers to his freedom – monarchy, aristocracy and established church – had been destroyed.

It should, however, be noted that this optimistic American ecumenicism, which was to provide the basis for the peace movements of the nineteenth century, was largely limited to the north-east of the United States. Further west and south, the experience of settlement, frontier defence and territorial extension was producing a war culture that owed nothing to an aristocratic feudal past and everything to the violent conditions of a frontier society. The heroic myths and imagery of the Wild West, as recreated in countless films, were eventually to dominate the American mind as strongly as had the myths of the Nordic sagas – the culture of the Norsemen – and the *chansons de geste* the mind of feudal Europe. It assumed no progress towards a peaceful global society, but a continual struggle in which the use of violence was justified by individual conscience and brute necessity rather than any sanction by church or state. It was understandably parochial, since keeping order within the parish left little time for wider considerations of global organization. A century or so later, when global organization began to appear possible and necessary, the image that came to many American minds was not that of balancing power between states, but of protecting law and order against its disturbers; the protection to be provided by a sheriff with his *posse comitatus.* If human corruption and inefficiency made this impossible, it must be provided by the efforts of a few good men following the dictates of a moral law within. This American populist belligerence, termed by some historians 'Jacksonian' after the first demotic president of the United States, sat uneasily with the irenical aspirations of the 'Jeffersonian' oligarchy, but it was long to outlive the feudal warrior culture of the Europeans.

*

This concept of an innate perception of a universal moral law also dominated the philosophy of that most remarkable child of the Enlightenment, the Prussian Immanuel Kant.

It was commonly accepted by the enlightened *philosophes* that men were naturally good but had been corrupted by institutions; and once those institutions had been reformed, natural virtue would reassert itself and mankind would live at peace. Kant believed nothing of the sort. Men, he believed, were built out of 'crooked timber', out of which nothing straight could be made. He agreed with his contemporaries that the immediate problem was that presented by the dominance of monarcho-aristocratic regimes for whom war was a natural and enjoyable way of life, and that the first step towards establishing peace should therefore be the establishment of what he called 'republican' states; not necessarily states where the monarchy had been overthrown, but where constitutions ensured that, before war was declared, there should be consultation with the people who would have to pay for and fight it. But this, though necessary, would not in itself be a sufficient guarantee of peace. War would still continue, warned Kant. But gradually its growing horror and expense would disincline peoples from waging it, and ultimately compel them to abandon the anarchical condition that prevailed among states and enter instead a 'league of nations', which would provide collectively the security that at present each sought individually. Further, all states should provide 'hospitality' for each other's citizens, a measure that would gradually create a sense of cosmopolitan community. The process would be a long one, with many setbacks; but what Kant called 'a seed of enlightenment' would survive all disasters, and ensure that progress would continue to the desired end. However improbable such an end might seem, he insisted, 'it is our duty to act according to the idea of such an end (which reason commands) even if there is not the least probability that it can be achieved'. It was a moral imperative to strive for peace, however remote might be the hopes of attaining it.

So if anyone could be said to have invented peace as more than a mere pious aspiration, it was Kant. He was almost alone in understanding that the demolition of the military structures built up in Europe over the past millennium would be no more than a preliminary clearing of the ground. New foundations would then have to be laid: peace had to be established. Its ultimate consummation would take a very long time, if indeed it ever occurred at all. Mankind was only at the beginning of what today we would term a very long 'peace process'. The fact that this process was to be ushered in by the most violent wars that Europe had seen for nearly two centuries would not have surprised Kant in the least.

III

PEOPLES AND NATIONS

1789–1918

It is no cause for wonder that once war became more popular, it should have become more violent. The *ancien régime* may have engaged in war as a matter of course, but those wars were kept limited by all manner of cultural constraints, not least their expense. The French revolution not only loosened the purse strings but released manpower on a scale that made it possible to field large armies, first of volunteers and later of conscripts, who were not trained, disciplined, paid or supplied on the scale of the old professionals. Lack of training made them rely on mass assault rather than disciplined firepower; lack of discipline and pay was compensated for initially by enthusiasm and later by hunger and ambition; lack of supplies meant reliance on foraging and loot. The size of these armies, and their capacity to dispense, however briefly, with regular supply lines made it possible to bypass fortresses; improvements made in road construction and cartography over the past century made their movements swifter, while similar advances in agriculture made it easier to live off the land. They learned to look for their reward not to a bankrupt government at home but to loot and plunder abroad. They were officered by men who had won promotion by talent, not birth, and led by a military genius. Napoleon Bonaparte knew how to turn to advantage not only all the apparent drawbacks of his new armies but also such technical innovations in the military art as had been taking shape in the last years of the *ancien régime*, especially the development of light and mobile artillery, whose fire could be used to supplement the assaults of infantry who through lack of training could attack only in huge, unwieldy columns. It was no accident that Bonaparte himself should have been an officer in the artillery, an unfashionable arm in which he had enlisted because he was not grand or rich enough to be accepted into the infantry or cavalry. It is an interesting example of social prejudice having an unforeseeable effect on the entire conduct of war.

What motivated these armies? Initially it was simple defence of their territory, as exhorted by the 'Marseillaise', against invaders aided and abetted by their own former rulers. This last factor was of enormous significance. The armies of France were not fighting for their king, but against him. But for whom *were* they fighting, and for what? Under the *ancien régime* the aristocracy had fought because it was a way of life, at the behest of a monarch to whom they had sworn personal allegiance. The professionals in the ranks had enlisted for money, and in battle were bound together, as soldiers always are, by group and regimental loyalty. The rest of the population – peasants, merchants, the educated laity, everyone included under the rubric the third estate – did not fight at all. Now they, too, had to fight; not just for their local community but, like the citizens of Marseilles, for the French nation that had taken shape in the early years of the revolution: a concept still remote for the inhabitants of what became known as *la France profonde*, but one that inspired the new elites in Paris and the provincial cities who had seized control of the machinery of the state and imposed their ideas, by force or persuasion, on the rest of the country. For them the nation was not simply a local tribe defined by geography and language, but a community that embodied the universal values of the Enlightenment. Like the founding fathers of the United States two decades earlier, the new rulers of France saw their nation as the standard-bearer of liberty, equality and fraternity for the whole of mankind. They had the duty to liberate men from their chains so that at last they could be brothers. Ultimately for the majority of the French populace this crusading zeal simmered down into the simple parochial concept of *la patrie*, that loyalty to country which their old adversaries the British had developed over the past century; but for decades to come the *tricoleur* was perceived throughout Europe as a threat to the existing order and a promise of a new one.

Initially the revolution found much support among the educated middle classes throughout Europe. Wordsworth for one, thought it bliss to be alive, while Goethe and Hegel never wavered in their support for Napoleon. Others thought differently, or, like Beethoven, changed their minds as events took their course. For one thing, soldiers are not always the best of ambassadors or missionaries, especially soldiers who, for lack of pay and supply, have to take what they need from the peoples they are supposed to be liberating. For another, Napoleon himself was a daemonic figure with his own agenda. His conquests certainly enabled him to destroy the old political order, sweeping away the relics of the Holy Roman Empire, rationalizing the governments of western Europe and introducing principles of law and administration throughout the continent that have lasted until our own time. But at the same time he transformed a republic that proclaimed the virtues of civic equality into an empire based on military hegemony, and France itself into a militarized society in which kingdoms were distributed as a reward for military success. Internal stability thus depended on continuous war. What was to be done with those vast armies if peace ever really broke out? So Napoleon mobilized against himself not only the governments of the *ancien régime* but the peoples they governed, and in so doing made them realize that they were peoples. Nothing creates a sense of national identity so quickly as having foreign soldiers quartered on your village. As a result, the Napoleonic era saw not only the development of organized warfare by land and sea, *la grande guerre*, on the largest scale ever seen, but also of *guerrilla* war; wars of peoples against occupying armies. This was to be no less significant for the future history, not just of Europe, but of the world.

Inevitably this universal doctrine of peace and brotherhood spread by French bayonets evoked an intellectual reaction. It accelerated the development of what has become known as the 'Counter-Enlightenment'; the view that man is not simply an individual who by the light of his own reason and observation can formulate laws on the basis of which he can create a just and peaceful society, but rather a member of a community that has moulded him in a fashion he himself cannot fully comprehend, and which has a primary claim on his loyalties. It was a view expressed in a mild form by Edmund Burke in England, and a profound one in Germany by Johann Gottfried von Herder. Both emphasized the uniqueness of their nation – for Herder and his German followers, the *Volk* – against the alleged universality of mankind; communities that based their claim to loyalty, not on abstract universal values, but on visceral sentiments more compelling than the demands of reason. This dialectic between Enlightenment and Counter-Enlightenment, between the individual and the tribe, was to pervade, and to a large extent shape, the history of Europe throughout the nineteenth century, and of the world the century after that.

The Napoleonic wars may have helped develop a sense of nationhood, but their influence on the spread of democracy was more problematic. The British, as we have seen, already had a clear concept of their unique nationality, one based more on xenophobia than on the brotherhood of man. For them the conflict was simply another round against the old enemy, France, and the fact that the French proclaimed the rights of man was, for many, a good enough reason for rejecting them. So far from advancing the cause of democracy in Britain, the quarter-century of warfare between 1793 and 1815 legitimized the oligarchic status quo, militarized the ruling classes and set back the cause of political reform by decades. The national hero was the ultraconservative Duke of Wellington, who fought successfully with an entirely unreformed professional army – the navy having saved the country from invasion and the traumas that might then have resulted. Elsewhere, Spanish peasants fought savagely against the French invader, but not out of any abstract loyalty to Spain, and under the leadership of the most reactionary church in Europe. In Russia and the Habsburg empire, loyalty to the existing regimes and dynasties was, if anything, strengthened, and their armies continued to make war in much the old style. Only in Prussia and the rest of Germany was there a widespread realization, especially among the educated classes, that their military defeat and political humiliation had been due not only to the incompetence of an army hitherto regarded as the best in the world, but also to their inability to match the commitment and enthusiasm that had inspired the armies of Napoleon. A German nation did exist, argued the nationalist thinkers Fichte and Arndt, but it had to be called into being, mobilized and led. That leadership could be provided only by the rulers of one of the most authoritarian states in Europe, the kingdom of Prussia. For the king and his advisers this remedy, fighting fire with fire, sounded worse than the disease, but they were overborne by their military experts. The king was prevailed upon to issue an appeal addressed to *meinem Volk*, 'my People', rather than to his 'subjects'. The Prussian army adopted as its motto *Für Gott, mit König und Vaterland* – God, king and country. The Germans, like the British and the French, now had a country, even if they did not have a state to match it. In 1813–14 the ranks of the Prussian army were swelled by volunteers and conscripts, though it is doubtful whether their presence made any difference to the conduct of the last months of the war. But was their loyalty primarily to *König*, or to *Vaterland*? The Hohenzollerns themselves were by no means sure.

*

Eventually Napoleon was defeated, and the rulers of Europe had the opportunity, unprecedented since Westphalia, of introducing another new world order. But there were three conflicting views as to what this should be: the Conservative, the Liberal, and the Nationalist.

The conservative view was held by the statesmen who

assembled to make peace in Vienna in 1814. For them the problem was twofold – to contain the power of France, and to limit the influence of the French revolution. Their leaders, Metternich, Castlereagh and Talleyrand, were as much the heirs of the Enlightenment as had been the French revolutionary leaders. They believed neither in the divine right of kings nor the divine authority of the church; but since church and king were necessary tools in the restoration and maintenance of the domestic order that the revolution had so rudely disturbed, their authority had everywhere to be restored and upheld. They therefore believed that international stability demanded the rejection of one of the basic tenets of the Westphalian system: the inviolability of the sovereign state. The great powers now claimed the right to intervene wherever international order appeared threatened by internal unrest. In fact, disagreement between them meant that the claim was exercised only briefly in the Italian and Iberian peninsulas in the immediate aftermath of the war, but nevertheless this breach in the Westphalian system was never to be entirely mended, and we see it crumbling again in our own day. It was a development that would have been highly unwelcome to Kant, who, having seen in his own lifetime the disastrous results of foreign interference in the internal affairs of

France, had made non-intervention one of the conditions for the establishment of peace. But this acceptance by governments of a common responsibility for the stability of the international system, even if it rested on no more than a desire for self-preservation, might be seen as an improvement on the international anarchy that Kant and his contemporaries so much deplored.

There was another positive feature of the Vienna Conservatives that distinguished them from their predecessors of the *ancien régime.* They no longer accepted war between major states as an ineluctable element in the international system. The events of the past twenty-five years had shown that it was too dangerous. Apart from anything else, the huge popular armies now required to fight major wars were themselves a threat to domestic stability. Wherever possible these were once more reduced to a hard core of loyal professionals officered by aristocrats, whose main function was suppression of revolution at home and abroad. The architects of the Vienna settlement still believed that major war was best deterred by a balance of power, and they tried to contain that of France by strengthening her neighbours, Holland, Prussia and Savoy. But we must note as another positive feature of the Vienna settlement that the preservation of peace was now seen as the joint responsibility of all the European powers, and the habit of frequent, if informal, consultation became established on a semi-permanent basis. The so-called 'Concert of Europe' that resulted fell far short of the 'League of Nations' visualized by Kant and his disciples, but it kept the peace not unsuccessfully for forty years, and then, after the turbulent 1850s and 1860s, for another forty. The Conservatives may have been concerned rather with the preservation of order than the establishment of peace, but in their eyes it came to very much the same thing.

The Liberals were understandably suspicious of the idea that preserving order was synonymous with the preservation of peace, seeing that they attributed to that order all the ills that had befallen mankind. They inherited their ideas from the British and Americans, rather than the continental European teachers of the Enlightenment; but like them they believed in the rights of man to self-government and self-determination, in the essential homogeneity and natural virtue of mankind, and in the inevitability of peace once the distortions imposed by the ruling elites were swept away. For them man's loyalty was not to monarchs but to mankind. Tom Paine, a writer equally influential on both sides of the Atlantic, was the prophet of this lay religion, and its believers were to be found among the increasingly powerful middle classes of Britain, the United States and France. They believed, with Adam Smith, Jeremy Bentham and later, most powerfully, Richard Cobden, that peace would be the natural consequence of the growth of international commerce and self-government, since it would increase the influence of those classes which, unlike the old ruling elites, had no interest in the perpetuation of war. Still marginal in continental Europe at the beginning of the nineteenth century, this group became increasingly significant in western Europe as the urbanization and modernization of society increased their political power, and improvement in communications bound them closer together across international barriers. By the beginning of the twentieth century the professional middle classes of the western world were indeed beginning to constitute what was later to be termed an 'international community'. Some indeed saw themselves as the international community: mistakenly, alas.

So whereas the Conservatives believed that peace consisted in the preservation of the existing order, the Liberals believed that it would result from a transformation of that order to be brought about by economic and social progress. The third group, the Nationalists, believed in an order based not so much on universal human rights as on the rights of nations to fight their way into existence and to defend themselves once they existed. In their eyes loyalty was due neither to monarchs nor to mankind, but to the nation. The latent opposition between Nationalists and Liberals was not immediately apparent. They were united in their hatred of the oppressive regime imposed on Europe by the statesmen of Vienna. Both initially supported demands for liberation: first for Greece, then for Italy, Hungary and Poland, and eventually for the more problematic claimants of the Balkan peninsula. But the nationalists did not expect that this programme would guarantee peace – certainly not in the short run. That would come only when all nations were free. Meanwhile, they claimed the right to use such force as was necessary to free themselves, by fighting precisely the wars of national liberation that the Vienna system had been set up to prevent.

It did effectively prevent them for over thirty years. The main threat to the Vienna settlement had seemed to come from a revisionist France that was not only chafing under an oppressive power balance, but remained the focus of revolutionary nationalism. In fact, that threat was wildly exaggerated. The French people had been exhausted by the Napoleonic experience. They now consisted overwhelmingly of peasant smallholders living on the confiscated lands of the church and aristocracy, who were as averse to war and alarmed by the threat of revolution as anyone else. But in Paris, the embers of revolutionary sentiment were still smouldering, fanned by Polish and Italian émigrés. The overthrow of the pacific Louis-Philippe in 1848 by what was little more than an urban *émeute* was widely seen as prefacing a rerun of the events of 1789–93. The French people themselves ensured that it did not. Even when Napoleon's nephew emerged triumphant from the turmoil three years later, his credentials were not so much those of Austerlitz as of Brumaire, presaging the restoration of order and the establishment of a generally acceptable domestic peace. Nevertheless, the events in Paris were taken as a starting gun by nationalist leaders all over Europe, and the precarious stability imposed by the Vienna settlement appeared shattered.

In fact, it was not. As both those shrewdest of political observers, Alexis de Tocqueville and Karl Marx, discerned, the bourgeoisie whose support was essential for any successful revolution took fright and aligned themselves with the forces of order – the old order. The leaders of that order realized that they could survive only by accommodating themselves to the forces both of liberalism and of nationalism, and exploiting them to their own advantage. Representative institutions of a kind were created throughout western Europe. Napoleon III exploited Italian nationalism to destroy Austrian power in Italy. Bismarck used German nationalism to destroy both Austrian and French influence in Germany and then stole the clothes of both liberals and nationalists by uniting Germany under the leadership of the monarchical and militaristic kingdom of Prussia. Austria bowed to the inevitable by granting autonomy to the kingdom of Hungary. By 1871, yet another new order had been created in Europe: that of nation states.

*

This new order, like all its predecessors, was created by war. The rapidly increasing pace of industrialization in Europe was hastening the emergence of a new structure of society in which power was based on industry rather than land, on cities rather than feudal estates. Germany had been unified by railways before it was unified by Bismarck, but its new economic and political potential had to be transformed into military effectiveness before it could change the balance of power. As Marx aptly observed, force is the midwife who expedites the birth of new orders, and in this case the midwife's instruments were those created by industrialization. Chief among these were the railways that could now deliver to the war fronts troops in quantities limited only by the size of the manpower available. Effective exploitation of railways demanded conscription of that manpower and the administrative skill to mobilize it; that is, a docile, if not indoctrinated, population and a highly efficient bureaucracy of a kind that Prussia already possessed and Austria and France did not. Simultaneously, the development of breech-loading firearms, and ultimately of high explosive, hugely extended the range both of infantry and artillery weapons, requiring the devolution of command responsibilities to ever lower levels; thus magnifying the importance, not simply of professional training and indoctrination, but of a high level of literacy among all ranks. War was becoming too serious a business to be left to an aristocracy whose inherent qualities of self-confidence and bravura, displayed to best advantage in the now highly vulnerable cavalry, were at best elegant appendages to machines that demanded highly qualified engineers to construct and maintain, and at worst recipes for spectacular suicide. Success in war no longer depended simply on the performance of armies on the battlefield, but on the administrative efficiency of the apparatus of state that put them there. Modernization was becoming an essential element in state survival, as was to be shown with terrible clarity in the Great War that broke out in 1914.

The new European order that was born in 1871 had much to be said for it. It provided satisfaction, at least in the short run, for all the three contending schools of thought described above. Obviously it was welcome to the nationalists, for whom the unification of Germany and Italy seemed to provide a Hegelian 'end of history', although there was still unfinished business further east. A similar satisfaction was felt at the same time in the United States, which had survived the threat to her nationhood posed by the civil war. It was no less welcome to liberals, for whom it brought the development of parliamentary institutions and facilitated the growth of a continent-wide infrastructure of trade and communications. As for the conservatives, whatever power they may have forfeited domestically, they retained a dominant position in the conduct of international affairs and were able to preserve through the second half of the century the systemic order, based on balancing power, that they had maintained in the first. The relative position of the players may have changed, Italy having joined the game and Prussia-Germany now dominating her previous rivals France and Austria, but a common interest in preserving the balance remained intact so long as Bismarck remained in power. The stability established in western Europe in 1871 was extended further east by the Congress of Berlin in 1878, where it preserved an uneasy peace in the Balkans for another quarter of a century. In perspective the twelve years of warfare between 1859 and 1871 might be seen as no more than necessary adjustments to unavoidable change, carried out with remarkable economy of force.

The forty-odd years of European peace that followed were not disrupted by the virtually continuous, low-level warfare being fought in Asia, Africa and North America to extend the area of European hegemony. This was driven by many factors other than the search for raw materials and markets which socialist economists saw as their principal cause. For the nationalists, it enhanced the greatness of their nation. For the conservatives, it provided employment for members of the old landed classes and useful practice for their armed forces. For the liberals, it was a *mission civilisatrice* and thus a moral duty. The disparity in standard of living between the West and the outside world that so hugely increased with the industrialization of Europe and the United States added to the 'white' man's sense of moral superiority, as well as providing him with the weapons and logistics that made expansion ever easier. A total lack of empathy and understanding of alien cultures, except on the part of a few specialists, made the conduct of colonial warfare all the more brutal. Massacres that were unthinkable in nineteenth-century (though not, alas, twentieth-century) Europe were taken for granted in the pacification of the extra-European world.

*

This new order seemed to work very well for another thirty or so years, and at the beginning of the twentieth century Europeans seemed to have every reason to congratulate themselves. So why, within a couple of decades, did the whole system collapse in the bloodiest war ever fought by mankind?

Historians have identified three underlying causes. First, there were the military dynamics at work during the apparent peace. The lesson learned from the Prussian triumph of 1870 was that victory went, in the words of the American general Nathaniel Bedford Forrest, to the side that could get there fastest with the mostest men. Whoever lost those first vital battles might be reduced within weeks to the category of a second-class power. The nations of Europe therefore converted themselves into armed camps. Their manpower was conscripted to military service at the age of eighteen and remained available well into early middle age to be recalled to the colours, loaded into railway carriages and sent into action. The dry, professional Helmuth von Moltke replaced the flamboyant hero Napoleon as the god of war. War itself ceased to be a romantic adventure and became a positivist science. The competitive modernization of weapons upset all attempts to maintain the balance of power and led to arms races that exacerbated international tensions and imposed ever higher demands on national budgets. The quest for security led governments to seek alliances to which they gave increasing priority over the maintenance of the Concert of Europe. Even for those who had no illusions about the horrors and complexities of twentieth-century warfare, defeat remained a catastrophe to be avoided at any cost.

Second, nationalism, in defiance of its prophets, did not make for international stability. A sense of national identity is not innate: it has to be taught and learned. As governments began to take more seriously the importance of mass education as an essential prerequisite for modernizing their states, minorities which had been left for centuries to speak their local dialects found imposed on them a language which they regarded as alien, even if it was that spoken by their ruling classes. German did not come naturally to the Polish inhabitants of West Prussia, or the Czechs in Bohemia. Magyar was not welcome in Croatia, any more than was English in Ireland. The Polish and Baltic inhabitants of the western provinces of the Russian empire increasingly resented attempts to 'Russify' them. Minorities began to explore their own ethnic and linguistic roots, recreated their own languages, claimed their own nationhood and, in consequence, their own independence. For the stronger states of the West, Britain and Germany, and for the ruthless autocracy of Russia, these remained largely internal problems, but they rendered the Habsburg empire increasingly ungovernable, intensified racial tension between peoples that began to identify themselves as 'German' or 'Slav', and internationalized the ethnic rivalries of the Balkan peninsula.

Liberals now found themselves in a dilemma. In conference after conference they issued condemnations of war, yet Serbs, Bulgarians and Albanians could hardly be condemned for fighting to free themselves from Turkish rule, as had the Italians to free themselves from the Austrian empire a generation earlier. For conservatives in the foreign ministries the matter was comparatively simple: their task was to prevent these struggles from upsetting the balance of power on which the peace of Europe still depended, so the less they interfered in each other's internal affairs the better. But how could the liberals of western Europe stand idly by while the Turks oppressed their European subjects with a mixture of indolence and massacre, and how could nationalists in Russia do nothing while their fellow Slavs were being oppressed, whether by Turks, Germans or Hungarians? In encouraging national self-determination, nineteenth-century Europe had opened a Pandora's box whose contents have not yet been exhausted.

Finally, in those western European states where national self-consciousness had been successfully consolidated by tradition or indoctrination, nationalism inevitably took a militaristic turn. History textbooks focused on military triumphs to be commemorated or defeats to be avenged. Military leaders became national heroes whose exploits were magnified by the popular press. As religious belief ebbed, nationalistic rituals replaced or absorbed ecclesiastical, and monarchs became icons of their nations. Conservatives and nationalists gradually amalgamated. Liberal internationalists, those loyal disciples of the Enlightenment, were certainly more active than ever in terms of the number of institutions and societies they founded, and the number of conferences they held. The Hague conferences of 1899 and 1907, with the consequent establishment of the International Court of Justice, were significant landmarks in the building of the 'international community'; yet even they abandoned their original objects of preventing war and building peace in favour of making war more humane for those actually fighting it. It was the tribalism of the Counter-Enlightenment that dominated the minds of the newly educated electorates. The original aspirations of Giuseppe Mazzini and his followers for world peace to be based on a union of liberated nations was degenerating into a kind of jungle morality, in which nations not only created themselves through war but showed their fitness to survive by fighting each other. War was perceived, by misinterpreters of Charles Darwin, as a necessary part of the natural order of things. In their view, peace led only to decadence, defeat and, ultimately, the disappearance of the peoples sufficiently misguided to pursue it. Hegel was mutating into Hitler; Mazzini into Mussolini.

It was this kind of nationalism that made war possible, even if it did not actually cause it; and it would be distorting reality to pretend that it was not stronger in Wilhelmine Germany than elsewhere in Europe. When war did break out in 1914, the peoples of Europe were ready for it: many welcomed it with enthusiasm, and all entered it with a clear conscience. The French fought to repel the German invader; the Russians, to rescue their Serb and French allies; the Austrians, for the preservation of their multiracial monarchy; the British, to fulfil their moral obligations, to uphold the balance of power and for the preservation of a status quo which they had comfortably dominated for the best part of a century. The Germans fought from loyalty to their only ally, the Austrians, and to repel the forces of Slav barbarism – but also, in the minds of some, to establish their place as a world power to which they felt that their strength entitled them; and finally, the Italians fought for the completion of the Risorgimento, the rescue of *Italia irredenta* still in Austrian hands.

Once they were in the trenches, the troops fought basically because they had to, and from group loyalty to their mates. But for those who thought about it at all, the appeal of country, or *patrie*, or *Vaterland*, united nationalists with those surviving conservatives who fought for God and king rather than country. And the concept of 'country' was not purely tribal. 'England' stood not only for the splendours of empire, but for democracy and the rule of law. 'Germany' was the home of a uniquely spiritual culture. 'France' was a historic civilization that united the two factions of republicans and Catholics still bleeding from the wounds of the Dreyfus case. And eventually for American liberals, like their British counterparts, it was a war which, by extending democracy, would itself bring an end to war and introduce a new international order.

These liberals had a good case. German nationalism did embody a unique strain of aggressive and expansionist militarism against which liberals within Germany itself had been struggling ever since 1871, and which coexisted uneasily in the Reichstag with a socialist majority that supported a programme of peace without annexations or indemnities. But that was not the policy of the German government, as was made clear in their September Programme of 1914 when victory seemed almost within their grasp; and although as the war went on the liberal opposition became increasingly bold and outspoken, so too did the demands of an annexationist right that drew its strength not so much from the old conservatives as from the lower middle classes and agrarian communities. The threat depicted by Allied propaganda, of an archaic militaristic rule of force backed by alarming industrial power that had to be destroyed if the liberal dream of universal democracy and the rule of law were ever to prevail, was not totally invalid. The new order that might have resulted from a German victory would have been a hegemony based, not like the Napoleonic on the concepts of the Enlightenment, but on military power untempered by any such universal values. The United States certainly had mundane reasons for entering the war in 1917, not least the heavy investment that she had already made in an Allied victory; but once she had joined, her explicit war aim was that of classical liberalism – the defence and extension of freedom and democracy. The message of the Stars and Stripes, like that of the *tricoleur* a century earlier, was one of liberation: the liberation not of a nation, but of the whole of mankind.

By 1918 the Europe that the Americans helped to liberate was no longer that of 1914. In the Napoleonic era social change had transformed the nature of war. Now it was the change in the nature of war that transformed the societies who fought it. The vast expense of waging industrial war had been foreseen before 1914, making some declare war to be impossible, others to believe that it could be kept short. It was on the latter assumption that the military made their plans, and they were proved wrong. On the eastern front, the sheer scale of operations made any decisive victory impossible in the short run. In the west, the increased power given to the defensive by the development of firepower without comparable improvement in communications and mobility led to complete tactical deadlock. After a year of slaughter, both sides abandoned their search for an operational decision on the battlefield and adopted instead a strategy of attrition, which demanded a total mobilization of their economies to provide war material, and of manpower to replace their battlefield losses. It was not so much operational failures on the battlefield as the sheer scale of the administrative problems of keeping their armies in the field, and in particular of their inability to feed their cities, that led to the internal collapse, first of the still only partially modernized Russian and Austrian empires, and then, despite its military and industrial strength, of Germany itself. France became economically dependent on Britain, and finally both on the United States. Thus it was that in 1918, with all the powers of Europe either defeated or exhausted, the United States appeared as a *deus ex machina* able to impose its own terms on allies and enemies alike. Once again a new world order seemed about to begin.

IV

IDEALISTS AND IDEOLOGUES

1918–89

In 1918 the Enlightenment liberals had, to all appearances, finally triumphed. The conservative order represented by ruling dynasties which could trace their descent back for half a millennium had disappeared overnight. Aggressive tribal nationalism, in the shape of Wilhelmine Germany, had been humbled. The United States, now the most powerful nation in the world, was liberal democracy incarnate, and its leader, President Woodrow Wilson, had a clear perception of the new world order he intended to introduce, together with the power to impose it. Or so it seemed.

Like Kant, Wilson believed that peace could not be established until all states were republican, so a prior condition of making peace was the destruction of the Hohenzollern dynasty in Germany; the subjects of the Romanovs and the Habsburgs had already taken matters into their own hands. Wilson also believed that the key to a peaceful world order lay in national self-determination, though by the time he left Paris he had come to realize that this was a great deal more complicated than he had at first assumed. Above all, he believed in the establishment of peace by the creation of a League of Nations who would mutually guarantee each other's security; one bound together not just by unwritten understandings based on perceptions of common interest like the old Concert of Europe, but by firm legal covenants: a vision sketched out by Kant and filled in by Bentham and his successors, that had become a central aspiration for the peace movements of the nineteenth century.

The League of Nations indeed seemed the fulfilment of all that the peace movements had fought for. Had all its members been like-minded, and had it included all the nations that mattered, the League might have stood a better chance. But they were not like-minded, and it was not comprehensive. Apart from the ostracism and humiliation of Germany, which in spite of the servitudes imposed by the treaty settlements remained the most powerful nation in Europe, it excluded a Russia whose leaders now offered an alternative vision of a world order; one that would come to have an appeal at least as attractive as that offered by the Western democracies.

Ever since the mid-nineteenth century, Karl Marx and his followers had been teaching that the French revolution was only the beginning of the gigantic process of social and political change that would result from the transformation of the basis of society from agrarian to industrial production. The nature and consequences of this process, they believed, could be scientifically analysed and predicted with as much certainty as Newtonian scientists could predict the movement of the planets. The bourgeoisie would, in time, replace the landowners as the ruling elite and create a global society based on the free movement of capital. This in turn would result in the accumulation of capital in fewer and fewer hands and the impoverishment of the masses who, uprooted from their native soil and alienated by the inhumanity of industrial production, would eventually rise and overthrow their exploiters. This revolution would lead, as had its French avatar, to an interim dictatorship provided by the Communist Party, but this would persist only for long enough to introduce a classless society. Then only could there be a just social order, and with it international peace.

In the miserable conditions of the early industrial revolution this combination of scientific certainty and millennial hope had seemed overwhelmingly attractive. But by the end of the nineteenth century the great bulk of social reformers in western Europe had accepted that the condition of the working classes could best be improved by working within the framework of parliamentary government that had been introduced throughout the continent, and their revolutionary wing had been reduced to a hard core of embittered ideologues mainly driven into exile. So long as the bourgeois order functioned effectively these ideologues were impotent. But in Russia economic development lagged behind that of western Europe by half a century. Bourgeois efforts at reform were punctuated by periods of savage repression, and revolution continually bubbled just beneath the surface. The impact of the First World War fatally undermined the entire Czarist regime, and its total collapse in 1917 enabled Lenin to establish the dictatorship of the Communist Party and proclaim his own new world order; or at least the promise of one when the world revolution had been completed. Meanwhile, as with the French revolutionaries of 1793, the new regime in Russia could not feel secure until it had liberated its European neighbours also; which, in 1918, it showed every intention of doing.

Thus already in 1918 two universalist concepts of world

order, both claiming the heritage of the Enlightenment, were contesting the future. Liberal democracy believed in the capacity of mankind, once liberated from historic constraints, to create orderly, just and peaceful societies through reasoned co-operation and agreement; while Communism put its faith in a historical process understood and interpreted by a disciplined secular priesthood, the Party, which had both the right and the duty to lead the struggle for a classless society, in the process destroying reactionary opposition and suppressing all dissidence within its own ranks.

*

In the event, the Versailles settlement provided no coherent plan for a new European order. Ironically, it was the very triumph of democracy that made it impossible. At Vienna Metternich, Talleyrand and Castlereagh had no need to take account of public opinion in reshaping Europe in accordance with their ideas. This indeed had been the main complaint of Bentham and his colleagues: a civilized public opinion, they argued, would be one of the major factors in the prevention of war. But one can never be sure that public opinion in democracies will be civilized, and in 1918 it certainly was not.

In France and Britain, wartime resentments were still too deep for the electorates (the British created for the first time by universal suffrage) to regard Germany as an acceptable partner in making the peace. The Germans themselves, irrespective of party, were resentful but not impotent, revisionist with the latent power to bring about a revision. The United States Congress was unready to accept any responsibilities as a member of the world community Wilson was trying to create, while the Russians turned their backs on it altogether. French statesmen, deeply sceptical of Anglo-Saxon liberal idealism, sought security by traditional means and tried to check German power by alliances: first with the British, who immediately defaulted; then with the new states of central Europe; finally with the Soviet Union. But in any case, French public opinion would support no serious military commitment to these allies. As for the British, where a war-weary public opinion was hostile to any European involvement, a liberal policy of conciliating Germany and preserving peace by appeasing her grievances became common ground for all political parties. When it came to the point, no democratic state was prepared to maintain the armed forces needed to support either an international order based on a traditional balance of power, or a rule of law under the auspices of the League of Nations.

In any event, the impact of the economic crisis of 1929, especially of the huge unemployment it created throughout the world, shattered belief in the old liberal certainties and encouraged support for the alternatives: Communism, with the Soviet Union as exemplar and leader; and, increasingly, Fascism.

Fascism should not be confused, as it is so often today, with right-wing authoritarianism. It was certainly authoritarian, but it was also populist and revolutionary; quite as hostile to bourgeois capitalism as were the Communists, even if prepared tactically to co-operate with it. Fascists had a very different idea of a new order. Their own roots lay deep in the soil of the Counter-Enlightenment; the belief in the community or *Volk* as against the individual, in intuition and emotion as against reason, in nationalism as against internationalism, and in will and action as against reasoned discussion and peaceful co-operation. They exploited the reaction against the boredom with peaceful bourgeois prosperity that had surfaced before 1914 in youth movements of every kind to find vivid expression in Italian Futurism, with its contempt for the past and belief in speed, violence and action as values in themselves. Its wider mass appeal was to xenophobia in general and anti-Semitism in particular; the Jews had always been seen as aliens in the midst of the tribe, as well as representatives of the international capitalism that was squeezing out petty-bourgeois entrepreneurs. Fascism, or rather national socialism, was to flourish most prolifically in postwar Germany, where the Movement, with its own hierarchy, symbols and rituals virtually replaced a traditional nationalism now associated with defeat or betrayal. Legitimacy for political authority was provided by the will of a leader who embodied the spirit and will of the *Volk.* Its concept of order was hegemonial and hierarchic, but the very term 'order' is too static to describe Nazi objectives accurately. The order was that of an army on the march. What mattered was the Movement itself rather than the goal: the struggle was its own justification. In so far as Hitler had a world vision it was one of a hegemony kept on its toes by continual conflict, a caricature of a Roman empire in which the rest of the world consisted either of subordinated associates or barbarian enemies. Peace did not exist in the Fascist vocabulary, except as a term of mockery or abuse. Fascists regarded war not just as an instrument of policy but as a thoroughly desirable activity in itself.

How could this be possible after the experience of the Great War? The memories of that conflict were ones of horror, and the prospect of another was regarded with dread. But contrary to general belief, the conduct of that war had not been completely sterile, nor were memories entirely negative. On the western front, in particular, there had been operational developments which were to have profound social as well as military consequences. There the deadlock resulting from the ascendancy of firepower over mobility had gradually been broken; not simply by the emergence of the tank, which was still too primitive to have much effect beyond that of surprise, but by the precise and flexible use of artillery in support of storm troops capable of providing their own firepower with light machine-guns, flame-throwers, grenades and mortars; groups in which command was devolved down to junior officers and below. To these developments on the ground was gradually being added a further element with enormous implications for the future of war: air-power. The vital element of radio communications needed to weld all this together had yet to be added, but enough was there to enable far-sighted strategic thinkers to visualize a new kind of war that would give scope both for professional skill and individual heroism. It would be waged, not by masses of conscripts commanded by château-bound generals far behind the lines, but by keen young specialists in violence: tank commanders, airmen, storm-troopers. The First World War had produced just such men, especially in the German army. Their symbol was the grim, purposeful *Stahlhelm* that had replaced the faintly comic *Pickelhaube* in 1916. They were classless, efficient, and above all they enjoyed fighting. Peace left them *désoeuvrés*, as it had the knights of the fourteenth century. They provided the nucleus of the Fascist movement in Italy and the Nazi movement in Germany. The image of violence they conveyed was a vital factor in enabling both movements to seize and retain political power.

Hitler used this threat of force to intimidate his adversaries, international as well as domestic. He used the fear of war in itself, especially of air bombardment, combined with the parallel fear of Communism, to persuade European statesmen either to give him the benefit of the doubt in the belief that, once German 'legitimate grievances' had been appeased, the aspirations of the peace movement could at last be fulfilled; or, more realistically, to settle for a German hegemony that now appeared inevitable. When war came in 1939, the brilliant campaigns of 1940 seemed to settle the matter. Although Britain was able to stave off defeat until more powerful allies found themselves compelled to enter the war on its side, Hitler's prophecy of a thousand-year Reich seemed only a hyperbolic exaggeration of what Europe had now to expect. But there was no prospect of this being a peaceful order. In Hitler's programme, the subjugation of western Europe was only a preliminary to the conquest of the Soviet Union to make possible the creation of a self-sufficient and racially pure Reich capable of holding its own against the rest of the world. Beyond that, Hitler was already making preparations for war with the United States. For him, war was not simply an instrument for creating a new order: it was the new order.

In fact, Germany's apparently invincible war machine was a less effective instrument than it appeared. *Blitzkrieg* only worked against weaker or unprepared opponents. Once Germany's adversaries had time to build up their own armoured and air forces, which the oceans provided for the Anglo-Saxons and their huge expanses of land for the Soviet Union, the pace of war slowed down. The deadlock of the western front was not repeated; but the naval and air superiority that made mobility possible, and the opportunities for armoured forces to make spectacular breakthroughs by land, still had to be fought for. All involved prolonged attrition. As in the First World War, the capacity to prevail depended not only on operational skills but also on social solidarity and successful war production. The difference now was that the production itself, and the civilians who were doing the producing, could be attacked directly from the air as well as indirectly by naval blockade. If there were no more Verduns and Passchendaeles (at least in the West; arguably Stalingrad was worse than either), there were now Coventrys, Hamburgs, Dresdens and, at the other end of the world, Hiroshimas. Ultimately the Allies were able to produce more ships, more aircraft, more tanks and, in the case of the Russians, more men. By the end of the war, the Third Reich resembled the Confederacy in the American civil war; fighting desperately and with great professional skill to the last, but being crushed by the application of hugely greater industrial strength.

On the Russian front, however, Hitler was fighting a kind of war that had not been seen in Europe for around a thousand years; one not simply for the defeat, but for the virtual annihilation of his adversary. His ultimate object was to destroy the Soviet Union as a political entity, occupy the territory of European Russia and colonize it with German peasantry to balance what he regarded as the over-industrialization of the Homeland. This would of course involve the elimination of the Jews, in implementation of his belief in the need for racial purity as the basis for a healthy society. In Hitler's view, this programme was no different from the way in which the Europeans had subjugated or eliminated the indigenous inhabitants of the American and Australasian continents, and there are certainly sinister similarities. But the Soviet Union, if still lagging behind the West, was now a modernized, literate industrial society, capable of fighting back with modern weapons on a scale that Germany could not match. In addition, the inhabitants of the overrun territories reacted to occupation with ferocious guerrilla warfare, which in the eyes of the Germans legitimized their own policy of annihilation. Simultaneously, such warfare was being waged in the Balkan peninsula, whose people had long experience of this sort of fighting against the Turks. Partisan warfare, as it came to be called, was not the least important legacy that the Second World War, like the Napoleonic wars, would leave to posterity.

*

In 1945 there seemed to be another chance to create a new world order, this time one truly global. Fascism had been not only defeated but discredited on its own terms: it had chosen the sword and perished by the sword. In the process it had become clear that military power was necessary not only to the *establishment*, but also to the *preservation* of peace. The unconditional surrender of the defeated powers meant that the victorious Allies felt free to reshape the world in their own image. For a few months it seemed that the foundations of peace had been established at last. The world would now be ruled by a consortium of democratic nations, freely collaborating within a constitutional structure based on that of Western parliamentary democracy. The General Assembly of the United Nations would provide its legislature and the Security Council its executive, with power to authorize armed action against disturbers of the peace. The vision, framework and muscle in building this new order was overwhelmingly that of the United States. Her European allies had their doubts. Archaic belief in the balance of power had not been entirely swept from their minds, if only because they still doubted the reliability of the new American commitment to upholding world order. For Britain and France, the main threat still seemed that of a revived Germany. The British built up French power as fast as possible to help contain it, and did little to discourage the Soviets from building up their own protective glacis in eastern Europe. The French under Charles de Gaulle, as mistrustful of British as of American commitment, attempted to revive the traditional entente with Russia. But it soon became clear that the Soviets had other ideas.

Unsurprisingly, Stalin did not share the liberal vision of world order. His own vision was probably one of a world from which capitalism had disappeared and the world's workers had united in perpetual peace under the direction of the Communist Party, which would in its turn take its direction from Moscow. But until that happened the struggle against capitalism would continue unabated, and the Soviet Union would have to survive in a world composed largely of hostile states. The power of the Soviet Union should, therefore, be extended as far as possible – certainly none of the hard-won conquests of the Second World War should be abandoned – and its capitalist adversaries should be weakened by continuing propaganda and subversion. The measures that the Soviets took to consolidate their power in central and eastern Europe – the elimination of internal opposition between 1947 and 1950, the Berlin blockade in 1948, the suppression of the Hungarian rising in 1956 and Czech independence in 1968 – may have been strategically defensive, but they were quite incompatible with a Western concept of world order based on national sovereignty and self-determination. Indeed, they bore a strong family resemblance to Hitler's actions in the 1930s, though perhaps a closer analogy would have been those of Metternich and Alexander I in the 1820s. The invasion of South Korea – for which Stalin certainly bore some responsibility, though not so much as the West believed at the time

– suggested that Soviet ambitions were indeed global. And so the Cold War began: a confrontation between two sides with incompatible visions of world order, each believing that peace ultimately could be established only by the elimination of the other. Fortunately, they came to appreciate that any attempt to do anything of the kind would probably involve the elimination of both.

*

The nuclear weapons dropped on Hiroshima and Nagasaki in August 1945, with their explosive power measured in kilotons and their victims numbering in scores of thousands, were initially seen by military experts as the twentieth-century equivalent of the machine-gun: a means of producing rather more economically the results that were already being achieved by more traditional weapons. All belligerents now regarded civilians as legitimate targets; whether directly for their role in war production, or indirectly because only their support, however unwilling, made it possible for their governments to continue fighting at all. The Americans had already killed more Japanese through the fire-bombing of their cities during the summer of 1945 than were to die at Hiroshima and Nagasaki. Had nuclear weapons not been developed and the Japanese government not sued for surrender when it did, they would no doubt have gone on to kill many more. The atomic bombs in fact achieved their results not so much through the physical losses they inflicted as through the effect, well-known on traditional battlefields, of shock. In spite of instant prophecies that their existence would now make war impossible, military planners realized, first, that the aircraft needed to deliver these weapons were themselves highly vulnerable; and second, that far larger numbers of bombs would be needed to have the same effect on a large continental power as they had on a crowded and urbanized island such as Japan. When confrontation developed between the Soviet Union and the West shortly after the ending of the Second World War, the military on both sides foresaw business as usual. The Soviets planned to advance their western glacis to the Atlantic to deny to the Americans the use of air bases in western Europe, while the Americans hoped at least to retain bases in the British Isles, Spain and the Middle East from which to bombard the Soviet Union and then, in due course, 'liberate' Europe for a second time.

The peoples of Europe knew nothing of these plans, and would have shown little enthusiasm for them if they had. 'Next time,' remarked a French prime minister, who did know about them, to his American colleagues, 'You will be liberating a corpse.' But within a few years such ideas were as obsolete as the Schlieffen Plan. Thermonuclear weapons were being developed with megaton yields dwarfing those dropped on Hiroshima and Nagasaki. Simultaneously, the introduction of intercontinental missiles supplemented, if they did not render unnecessary, the use of manned aircraft to deliver them. Half a dozen such missiles would be more than enough to destroy the United Kingdom; two dozen, the United States. Civilian populations, which at the beginning of the century had been regarded as reservoirs of military manpower and in its middle years as producers of the tools of war, were now no more than hostages. Wars between states might continue, but – at least in the developed world – they could no longer in any sense be regarded as wars between peoples.

It has often been said that between 1945 and 1989 peace was kept by a war that nobody dared to fight. But even if nuclear weapons had not been developed, war would not necessarily have broken out in Europe, except perhaps through some ghastly miscalculation. War-weariness was far too widespread. Once the basis for Soviet support in western Europe had been neutralized, partly through the effectiveness of overt or covert American aid and partly through revulsion at Soviet actions behind the Iron

Curtain, Stalin was far too shrewd to have attempted overt aggression, and his successors were even more cautious. As for the West, in spite of the rhetoric, no one was going to start a third world war to liberate its *irredenta* behind the Iron Curtain. Indeed the west Europeans used the American nuclear umbrella as an excuse for running down the armed forces for which their increasingly wealthy and pacific electorates were as reluctant to pay as they were to use. West Europeans, indeed, freely accepted American military hegemony as a cheap way of protecting their own security. The North Atlantic Treaty Organization gave the British a privileged position within the alliance; it hastened the German and Italian restoration to international respectability; while the French, however much they tried to dissociate themselves from the servitudes of the alliance, had nowhere else to go. But basically American dominance was accepted in western Europe because it guaranteed an order a great deal preferable, either to the Fascism its inhabitants had so painfully escaped or to the Communism they could see in action beyond the Iron Curtain. The West, with all its historic differences and rivalries, was now beginning to constitute a real cultural community.

*

During much of the Cold War, Western public opinion was divided between three schools of thought. First, there were those who had no doubt that the war was a real one and were determined to win it, preferably without the use of nuclear weapons but if necessary with them: a group rather more numerous and vocal in the United States than in western Europe. Second, and probably in a large majority on both sides of the Atlantic, were those who accepted the existing situation with all its imperfections as the best peace they were likely to get, and were primarily concerned with stabilizing it through deterrence. Finally, there were those who saw nuclear deterrence as a threat rather than a safeguard, and campaigned vigorously for its abolition. Between them they engendered a quite enormous body of literature, but a brief word about the last seems to be in order.

By the end of the 1950s this peace movement had gathered considerable support, exploiting the very natural and widespread fears that the peaceful resolution of the 1962 Cuba missile crisis had done little to allay. In its European branches there was a strong element of resentment at American dominance, but it was increasingly fuelled on both sides of the Atlantic by a more generalized rejection of the capitalist system in all its manifestations; a rejection that had by the end of the 1960s become a major cultural phenomenon throughout the West. The cruciform symbol of the British Campaign for Nuclear Disarmament came to represent an entire counter-culture on both sides of the Atlantic, for whose members peace was not to be found in a world which they saw dominated by the military–industrial complex and based on injustice and terror, but in a future where all arms had been abolished. This movement did not sympathize with the Soviet alternative: it ignored it. While rejecting what, with some reason, it saw as an American world hegemony, it did not find it necessary to formulate an alternative world order. As with the protests against the arms race that had been an intrinsic part of such movements before both world wars, nuclear, just as conventional, weapons were seen as the cause rather than the result of international tensions. The assumption was that if only the West would cease threatening the Soviet Union with nuclear weapons, the Soviets would reciprocate, and all problems could then be peacefully resolved under the auspices of the United Nations. Driven by moral fervour, it did not overly concern itself with prudential considerations of power.

In fact, the confrontation between the two world orders neither erupted into overt war nor ended with unilateral disarmament, but it softened with time. The Cuba missile crisis reminded both sides of the appallingly high stakes for which they were playing. A more pragmatic generation came to power in the Soviet Union concerned more with resolving the contradictions of Communism than with exploiting those of capitalism. In the United States, the crusading zeal so notably expressed in President Kennedy's inaugural address was dispelled when the experience of the Vietnam conflict showed what 'paying any price' and 'bearing any burden' actually meant. Then in 1968 a new approach to international relations was brought to Washington by the appointment of Dr Henry Kissinger, first as National Security Adviser and then as Secretary of State.

Kissinger, a first-generation European immigrant and himself a notable historian of European diplomacy, had little time for the idealism that inspired so much of American foreign policy. He took as his models the statesmen who had remoulded Europe after the Napoleonic wars. Whereas then the great powers, whatever their other differences, had a common interest in preserving peace and preventing revolution, they now had a similar interest in co-operating to avoid nuclear war. In his view, traditional national interest was a better guide for statesmen than ideological commitment, and peace was best preserved by old-fashioned diplomacy working to maintain an acceptable balance of power. Such a policy was made easier by the emergence of the People's Republic of China as an independent power equally hostile to the Soviet Union and to the United States. Kissinger found the leaders of both Communist powers to be men after his own heart, concerned as they were more with the realities of power than the promises of ideology, and he was able to establish with them relations rather warmer than those he enjoyed with many of his own colleagues in Washington. But such relations could be conducted only with a secrecy that, while it had presented no problems to the statesmen at Vienna and even fewer to Kissinger's interlocutors in Moscow and Beijing, was made possible in the United States only by the support of a president, Richard Nixon: a man who already saw himself in a state of war with the rest of the Washington establishment, and whose paranoid attempts to cover his traces led to his disgrace. Once again, the government of the United States had to accommodate itself to the complex and contradictory tides of public opinion; and as we have seen, public opinion is not necessarily so civilized as Bentham would have wished.

Kissinger had further hoped to soften the stark confrontation of the Cold War by building a multipolar system comparable with the old Concert of Europe, in which the world would be governed by a consortium of the United States, the Soviet Union, Japan and Europe. Multipolar systems, explained American political scientists, were inherently more stable than bipolar systems. Be that as it may, historical paradigms could not be so easily resurrected. In spite of her new-found wealth Japan was too timorous to assume such global responsibilities, while Europe, if now rather more than a geographical expression, was still a long way from developing the unity required by a common defence and foreign policy. Both were still creatures of the United States, in the literal sense of the word. But the individual states of Europe were not. The British still followed in the American footsteps like a loyal, if sometimes bewildered, gun dog, but the death of de Gaulle had not substantially altered the determination of the French government to distance itself from American policy on every possible occasion. As for the German Federal Republic, although remaining a loyal alliance member, it had begun to pursue its own independent *Ostpolitik* aimed at peacefully eroding its barriers, with eastern Germany in particular, and eastern Europe in general. This was to bear fruit in the Helsinki Accords of the 1970s, whose stipulations for free communication, trade and political co-operation across the Iron

Curtain went a long way to establishing peace in the Kantian sense.

*

But Europe, if still the most important, was now by no means the only region in the world where peace had to be established. For the best part of a century the world order had been the European order writ large. Most of the world outside the Americas and China had consisted of formal or informal European empires. European peace was, by and large, world peace, while European wars had been world wars and had been so since the eighteenth century (the so-called 'First' World War was in fact about the sixth). But with the Second World War, the will and capacity of the Europeans to maintain their imperial hegemonies were exhausted. If there was one thing on which the Americans and the Soviets were agreed, it was in greeting this development with wholehearted satisfaction. Washington initially regarded the emergence of the post-colonial states with as much benevolence as it had the new states of Europe in 1918–19. But when they emerged under Communist leadership, or at least with Soviet support, declared themselves non-aligned in the confrontation of the Cold War and refused to accept American leadership in the United

Nations, the benevolence rapidly waned. Cold War imperatives came to the fore. Loyalty to the West became the condition of approval, and the United States generously supported any government that professed it, however authoritarian, with arms, aid and technical advice. The Soviet Union did the same with their own clients, but far less effectively – except in the field of social control where they excelled. It has to be said that even where there were genuine efforts to install democratic institutions on the Western model, as the British attempted throughout their former African possessions, these rapidly failed. It seemed that stability in these societies could be provided only by authoritarian rule; usually by a Western-trained military which, it was hoped, would develop the economy and strengthen the small educated elites on which the successful running of the state would depend. Once again, it became clear that democratic institutions are not in themselves sufficient to produce democracy in cultures which, whatever their own merits, have not developed the kind of civil society in which the Western model was rooted. If many Third-World leaders ultimately sided with the West, it was less from ideological preference than because it was made very worth their while.

The regional conflicts that erupted throughout the

world in the wake of the withdrawing European hegemonies were widely seen as mere proxy wars between the Soviet Union and the West. But fundamentally they were the kind of struggles for power, local or regional, such as always followed the collapse of empires. Although exacerbated and sometimes prolonged by the Cold War, they would have occurred even if the Cold War had never happened, and were to continue long after it had ended. In some regions, such as the Indian subcontinent, they were to be fought with the elaborate weapons of the developed world. Elsewhere, insurgents, whether against the old colonial regimes or their successors, imitated the techniques developed by Mao-Zedong against the Japanese in the 1930s and the Chinese nationalist government in the 1940s, enlisting popular support by a combination of propaganda with guerrilla warfare, and building up the new order as they demolished the old. It was a technique that the West – and in its turn the Soviet Union – found difficult to counter, whether it was used against their own forces or those of their surrogates. Western conscript armies, over-armed and under-motivated, fought in alien environments surrounded by populations in which they found it difficult to tell friend from foe. Success demanded a degree of patience, military skill, political sagacity and, above all, domestic support that was not forthcoming from Western electorates now overwhelmingly concerned with their own welfare and whose elites were deeply divided over the wisdom and morality of the whole enterprise. On the whole, events were to justify the sceptics. There were many ways in which the West could influence the development of the post-colonial world, but, with very rare exceptions, overt military intervention was not one of them.

*

The Cold War ended with dramatic suddenness in 1989, and historians will long debate why. There is no doubt that the West 'won', and that there was a strong military element in that victory. Clausewitz had pointed out that perception of the outcome of an engagement might make it unnecessary to fight one at all. During the Cold War, military policy on both sides had been directed to persuading the adversary that if he did go to war he would lose; if not by being defeated in a traditional sense, then through provoking a nuclear catastrophe in which such terms as 'defeat' and 'victory' were irrelevant. This involved not so much a numerical arms race, as keeping abreast of rapidly evolving weapons technology at huge expense. The old tag, *pecunia nervus belli*, applied as much in Cold War as it had in hot.

The United States could set the pace at acceptable cost to the national economy. The Soviet Union could keep abreast only by starving the civilian sector and slowing down or even reversing the improvement in a standard of living of which, in the 1950s, it had been quite justifiably proud. Eventually in the 1980s, perhaps partly as a result of Reagan's visionary 'star wars' project (which, whatever its practicability, indicated the huge lead the Americans enjoyed in military capabilities) but more generally with the advent of a whole new era of information technology with far-reaching implications for combat effectiveness, it became clear to the Soviet leadership that they could sustain the military balance only at unacceptable cost. In consequence they took the quite amazing decision that the confrontation should if possible be brought to an end.

The decision may have been amazing, but it was entirely logical. In spite of its undeniable achievements in modernizing Russia, the Communist system had not delivered on its promises after three generations of sacrifice, while capitalism had flourished like the proverbial green bay tree. Fascism had failed because it did not deliver the promised military victory; Communism, because it failed to deliver the promised prosperity. It was no longer possible to keep the Soviet people in happy ignorance of what was happening beyond the Iron Curtain; certainly not the peoples of central and eastern Europe once the Helsinki process had got under way. Their massive peaceful protests were to expose the hollowness of the Soviet hegemony. Mikhail Gorbachev had hoped that the peace established at Helsinki and confirmed by the treaties of 1990–91, which reunited Germany and liberated eastern Europe, would be compatible with a reformed Communism, but he was wrong. The Party disintegrated, and, save for a few rock pools such as Cuba and North Korea, Communism vanished as an alternative ideology to that of Western democracy. For the third time in a century, a president of the United States was able to proclaim the inauguration of yet another new world order.

V

TOMAHAWKS AND KALASHNIKOVS

AD 2000

In the last decade of the twentieth century the liberal inheritors of the Enlightenment seemed once again poised to establish peace. The prospects now seemed brighter than ever. There was none of the confusion and conflicting aims that had prevented a genuine settlement in 1918; none of the *de facto* division of the world that had existed in 1945. No alternative model for a world order was on offer: that of Kant and his disciples seemed to have triumphed over all its competitors. So far from resenting this, the countries of the former socialist commonwealth competed to join it. Even the People's Republic of China, while retaining the disciplined structure imposed by the Party, were using it to modernize the country as fast as possible in order to compete in the new global market economy. The consensus seemed cemented by the challenge from Saddam Hussein, a local warlord anachronistically using force to create a regional hegemony, against whom the United Nations came together under US leadership in a show of collective security considerably more credible than that evoked by the Korean challenge in 1950. At last the world seemed free in the sense understood by the men of the Enlightenment: free for travel, free for the communication of ideas, free above all for commerce. One enthusiastic American analyst suggested not only that a new world order had begun, but that history itself had come to an end. His European colleagues, who had some experience of history's capacity to pick itself up off the floor and deliver powerful blows to the solar plexus, were rather less sure.

Their scepticism proved justified. Within a decade, the general mood had turned sour, and the new millennium was to be greeted with apprehension rather than hope. For this there are two fundamental reasons.

The first is that freedom of communication, above all freedom of commerce, does not necessarily establish peace. Marx's warning of the tensions created by a global market economy remain valid, even if his prescription for resolving them has not. In a global capitalist society, as in a domestic, the weakest goes to the wall. Global competition often results in local ruin. Already in the nineteenth century infant industries and long-established agricultural producers had pestered their governments with demands for protection against their industrially more advanced competitors. Resentment against the power of international capitalism had been particularly intense in the less developed countries of eastern Europe, where it produced a popular backlash everywhere between Moscow and Vienna: a backlash not so much, as Marx had expected, among the proletariat, as from the petty bourgeoisie of the cities and small farmers in the countryside, feeding the lethal mixture of xenophobia, anti-Semitism, nationalism and authoritarianism that had ultimately, as we have seen, ripened into Fascism. Between the wars both Fascist and Communist states had created command economies to protect themselves against the fluctuations in international markets that had such catastrophic results in 1929, and their apparent success had been widely envied in the West. The economies of the Axis powers had been destroyed in the Second World War, but their recovery was aided by lavish help from the United States. In any case Germany and Italy had powerful entrepreneur classes and Japan a strong corporate culture that helped them make the transition to market economies.

But the demolition after 1989 of the protective barriers that had enabled the socialist economies behind the Iron Curtain to survive and moderately prosper created only economic confusion and widespread misery. Optimists have seen this as a transition period, comparable with the early days of capitalism in the West; but in fact what has eventually emerged in the former Soviet Union, as had already been the case in much of Africa, is not so much democracy as 'kleptocracy'. Capitalism, or the rule of the market, is effective only when practised by communities where there already exist stable civil societies held together by efficient bureaucracies and common moral values, conditions that the market itself is powerless to create. Democratic elections have often had the effect of destroying such social cohesion as already existed. Under these circumstances there has tended to emerge in post-Communist societies not the order and prosperity that liberals have hoped, but the kind of xenophobia, authoritarianism and nationalism that had spawned Fascism a generation earlier. A pessimist might conclude, as did William Golding in his disquieting fable *The Lord of the Flies*, that this is the only condition really natural to mankind.

In the second place, there remain many parts of the world where Western values, and the whole process of economic modernization associated with them, are regarded as culturally alien, a threat to indigenous values and social cohesion. In these regions protest is strongest and best organized among established religions whose leaders see themselves as the guardians of the traditional order, as did the Catholic church in Europe throughout the nineteenth century. Certain sects in Islam, especially in Iran, provide the obvious examples. In such cultures a visceral hostility can be mobilized against the principal *Kulturträger* of the new order, the United States, much as a comparable hostility had been mobilized in nineteenth-century Germany against France as the *Kulturträger* of the revolution. The Stars and Stripes, like the *tricoleur* before it, is seen as the symbol not of liberation, but of alien oppression. But resentment of modernization can be found within fundamentalist Christian sects as well, not least within the United States itself. The conflict is often as much domestic as international, between urban and rural communities within the same state. In many Third-World states authoritarian regimes have been able to seize and retain power by mobilizing the populist backlash against the modernization led by Westernizing urban elites or ethnic groups. Elsewhere a loosely knit international organization of extremist *groupuscules*, inspired by the kind of religious fanaticism that had long disappeared from the modernized West but armed with the latest of modern weapons, have dedicated themselves to the overthrow of the American-led secular world order and to the establishment of a kind of theocracy not seen in Europe since the early Middle Ages.

The situation has been further complicated by the question mark that now hangs over the very essence of the Westphalian world order: the sovereign state itself. This finds itself increasingly vulnerable to three different kinds of corrosion – from above, laterally and from below.

Gradually, under the pressure of globalization, supranational entities are coming into being with the power to override the authority of sovereign states. Initially these were created to handle economic questions, but increasingly they have extended their jurisdiction, in order to create level playing fields, into social legislation as well. This process is most marked in the European Union, which is, at the time of writing, trying to dictate to one of its members (Austria) how its own elected government should be composed, but one cannot draw too far-reaching conclusions from this model for the rest of the world. The United States in particular has the will and the capacity to resist such intrusions into its own sovereignty almost indefinitely. But the weaker the state, the more its sovereignty is likely to be challenged. Recently the self-denying ordinance in the UN Charter against interference in the domestic affairs of its members has been set aside, specifically by the United States and its closest European allies, who claim an overriding right of forcible intervention in the affairs of sovereign states to prevent the abuse of human rights, whether or not that intervention is sanctioned by the United Nations. There is nothing particularly novel about this claim. It bears a strong family resemblance to the arguments that throughout the nineteenth century led the European powers collectively to intervene in the affairs of the Ottoman empire, as well as individually elsewhere in the world as a preliminary to bringing those barbaric regions within the pale of Western civilization. A comparable right of intervention in financial affairs was then asserted in cases of chronic indebtedness – something which the activities of the International Monetary Fund have made familiar in our own times. The universal recognition and, where necessary, the enforcement of such humanitarian norms would once have been universally regarded as a mark of human progress, but there are those even in the West who now consider it a manifestation of cultural imperialism. Certainly, few states can be considered effectively sovereign if their more powerful neighbours can exercise this kind of *droit de regard* over their domestic affairs.

Perhaps yet more significant are the lateral pressures of globalization. From the days of the East India Company to those of Cecil Rhodes in South Africa, powerful foreign business interests have been able to bypass or control the governments of weaker states. The capacity of multinational corporations to do so today may have been much exaggerated, but not even the most powerful economies are immune from their pressures. Even that most basic attribute of sovereignty, control of the currency, has been eroded if not entirely destroyed by private manipulation of financial markets. In addition, the global explosion of communications has strengthened the hand of private pressure groups, non-governmental organizations dealing with environmental, ecological and humanitarian matters. Sometimes so-called sovereign administrations seem no more than battlefields over which these non-state actors carry on their conflicts.

But the most important source of erosion of state authority has been the diminution of its support from below. It has to be said that a high proportion of states that have come into existence since 1945 have not developed as nations at all, for one very basic reason. They have not experienced that essential rite of passage: fighting, or at least showing a credible readiness to fight, for their independence. Their colonial masters, after learning a few sharp lessons, left them swiftly and peacefully to their own devices, often helpfully providing them with the tools for nation-building before there was any nation to build. When fighting for liberation did take place, it was often only by small, dedicated groups who had more support from sympathizers in the West than from among their own populations, and who thus found themselves after liberation involved in continuous civil wars. The populations of such states feel no particular loyalty to governments which are often engaged in unpopular modernizing programmes; deeply in debt to the West; thoroughly corrupt; or all three.

Even in fully developed industrial and post-industrial states, the waning probability of major war – certainly of war involving the efforts of the entire population – has eliminated the need for the kind of social mobilization that, as we have seen, brought states into being in the first place and sustained their cohesion through the traumatic experience of modernization in the nineteenth century. The intense popular commitment that had made possible the wars of nations and the building of those nations themselves was predicated on an individual commitment, however notional, to dying for one's country, or at least identifying oneself with, and supporting, those called upon to do so. War was the great audit of national efficiency and morale. It tested the claims to pre-eminence of ruling elites and provided the hidden agenda for the education of adolescent males of all classes well into the twentieth century. The symbols and rites of nationhood cemented social solidarity in an era when it was under its greatest stress. But by the end of the twentieth century death was no longer seen as being part of the social contract. War, or the ever-present possibility of war, no longer provided the cohesive force that held society together, and nothing comparable had emerged to take its place. Where conscription survived at all it was for social and political reasons. Militarily, it was increasingly counter-productive. War was now fought by highly trained specialists, as it had been in the age of the Enlightenment; specialists increasingly difficult to recruit and retain.

Under these circumstances the kind of patriotism that enabled the peoples of Europe to endure two world wars now appears as archaic as the feudal loyalties that it had displaced. The national flag is no longer a symbol to evoke awe and respect. At best it is the logo of a firm – Britain plc – whose function is to provide dividends for its shareholders. Those who, like Burke before them, lament the debasement of the nation into a joint-stock company are now hard put to it to find functional justification for their complaints. The state now has to justify its survival in the face both of the larger entities that seek to absorb it and the smaller that try to divide it, by the purely pragmatic argument that it can provide a better deal for its shareholders than either. Even its weapons are designed and produced with an eye as much to profitable export as to national defence. Indeed, European states now spend no more on their armed forces than is necessary to persuade the United States that they are *bündnisfähig*, capable of functioning as useful allies.

These generalizations do not of course apply to the United States, much less to Russia and the People's Republic of China. For the Chinese, the armed forces remain an intrinsic element in nation-building, while for the Russians, they may well provide a motor for national regeneration. But the United States is a more complex matter. American military supremacy over the rest of the world is now as great as that exercised by the European powers collectively a century ago. It provides the unchallengeable basis for the new world order. American elites remain very conscious of imperial responsibilities that are based on much the same combination of moral commitment and material interest that guided their European predecessors. The American people display neither the deference that characterized European populations a hundred years ago nor their visceral aversion to war today. For the Americans the flag is a great deal more than a logo, and woe betide anyone who treats it as such. A Jacksonian bellicosity remains very much part of American popular culture. But with it goes the reluctance common to all Western urbanized societies to suffer heavy losses, either civilian or military. Every serviceman killed in action is regarded as a national martyr, to be brought home and interred with full military honours. Every airman shot down in combat can rely on vast resources being devoted to his rescue. This paradox of what has been termed the post-heroic age has been only partially resolved by smart technology that makes possible the accurate destruction of enemy targets at great distances. Peoples who are not prepared to put their forces in harm's way fight at some disadvantage against those who are. Tomahawk cruise missiles may command the air, but it is Kalashnikov sub-machine-guns that still rule the ground. It is an imbalance that makes the enforcement of world order a rather problematic affair.

*

On the face of it this erosion of the nation state, if not of the state itself, might appear something that good Kantians should welcome: it was, after all, only the ideology of the nation state that made possible the terrible wars of the early twentieth century. The complaint often made against both the League of Nations and the United Nations is that these were leagues of *states* and not of *peoples*; and that if only peoples could get together behind the backs of their governments, they could at last establish peace. Exactly this, it may be argued, is what is happening today with the growth of transnational communications and the growth of the world pressure-groups described above. Is this not a wholly desirable trend?

The answer is yes, of course, as we shall see in a moment. But the trouble is that the state not only makes war possible: it also makes peace possible. Peace is the order, however imperfect, that results from agreement between states, and can only be sustained by that agreement. It is not clear what alternative creators and guarantors of peaceful order could or would take the place of the state in a wholly globalized world. The state still remains the only effective mechanism through which people can govern themselves, and it reaches the limits of its legitimacy at the point where its inhabitants can no longer accept it as representative of their community. When that point is reached they can only create new states. The erosion of state authority is thus likely not to strengthen world order but to weaken it, since states become incapable of fulfilling the international obligations on which that order depends. Supranational authorities can be effective only if they can inherit the loyalty that states can at present demand from their citizens, and that requires a homogeneity of culture and values that may take generations to develop. Even then, conflicts of regional interests can still destroy the entire structure, as was very nearly the case with the American civil war.

Kant was therefore quite correct when he saw the existence of republics as the essential foundation for the establishment of peace, but because such states can command the obedience of their citizens rather than because those citizens are necessarily averse to war. That obedience indeed made it possible, even in Kant's own time, to wage wars far more terrible than the limited wars of monarchs, but it was still the only basis on which a stable peace could be built.

But this in itself is not enough. Peace, as we have seen, is not an order natural to mankind: it is artificial, intricate and highly volatile. All kinds of preconditions are necessary, not least a degree of cultural homogeneity (best expressed through a common language), to make possible the political cohesion that must underlie a freely accepted framework of law, and at least a minimal level of education through which that culture can be transmitted. Further, as states develop they require a highly qualified elite, capable not only of operating their complex legal, commercial and administrative systems, but of exercising considerable moral authority over the rest of society. Where these conditions do not exist, or where they have decayed, there may well be no community of interest in creating, or capacity for sustaining, a peaceful international, or indeed domestic, order. Armed conflict becomes highly probable.

The establishment of a global peaceful order thus depends on the creation of a world community sharing the characteristics that make possible domestic order, and this will require the widest possible diffusion of those characteristics by the societies that already possess them. World order cannot be created simply by building international institutions and organizations that do not arise naturally out of the cultural disposition and historical experience of their members. Their creation and operation require at the very least the existence of a transnational elite that not only shares the same cultural norms but can render those norms acceptable within their own societies and can where necessary persuade their colleagues to agree to the modifications necessary to make them acceptable.

The role of such a transnational elite was filled in

medieval Europe by the church. The clergy spoke a common language, shared a common culture and belonged to an international order whose norms had both to be effectively transmitted to a local level and to be modified to take account of variations in indigenous cultural practice. Their ascendancy ended with the Reformation, but a new elite then emerged; that of that of the much intermarrying noble families who conducted the affairs of the emerging states, increasingly assisted by the literate, university-trained lawyers and scholars who were to provide the seedbed for the Enlightenment. Servants of their sovereign princes, they did not suffer from the conflict of loyalties that had destroyed a Becket or a Wolsey, but they did provide the framework that made it possible for an international system of states to function and for peace to be, however intermittently, established. Perhaps yet more important was that transnational community of bankers and merchants, quietly carrying on their businesses across frontiers often indifferent to the existence of war or peace.

Until the eighteenth century these people constituted a tiny minority in their own still largely illiterate societies, but one treated with considerable deference and regarded with great respect. In the nineteenth century as industrialization improved communications and Europe became urbanized and modernized, their numbers and importance increased. Diplomats and bankers were joined on the international stage by jurists, parliamentarians, entrepreneurs and professional persons of every kind. Command of foreign languages – usually French, but increasingly English – became a necessary professional skill. The community of scholars expanded to include scientists and technologists. International organizations and conferences proliferated. The modernization of Europe, although it had the immediate effect of encouraging nationalism, at the same time made its members increasingly interdependent. Without transnational transparency and co-operation, the professional classes could not provide the expertise that enabled national economies to function.

It was to this growing transparency and interdependence that liberals like Richard Cobden had looked for the eventual creation of a peaceful international order. But it took time. It was from among the ranks of these businessmen and professionals that the peace movement had been very largely recruited, and their influence in Protestant Anglo-Saxon nations was particularly strong. But elsewhere, especially in the newly formed German empire, these bourgeois *arrivistes* assimilated themselves to the older, still militaristic ruling classes; partly through a common fear of socialism, but very largely for reasons (never to be underrated) of social prestige. It was not only in Germany that they willingly provided the bulk of the officers needed in the new mass armies. Nevertheless, as the twentieth century wore on, their interaction and interdependence increased, and with it the strength of their common supranational interests. Simultaneously they were becoming increasingly vital to the prosperity and development of their own countries. The evisceration of the professions through the purging of their Jews fatally weakened Germany in the conduct of the Second World War, while the wholesale destruction of the bourgeoisie as a class by the Soviets in the 1920s, and the Chinese in the 1970s, was to cripple both countries in their race for modernization. If by the end of the twentieth century the United States was the most powerful nation on earth, it owed its primacy not only to its size and resources, but to its readiness to accept and adequately reward scientists, technologists, engineers and entrepreneurs from throughout the world and turn them into Americans.

So by the beginning of the new millennium there has come into being a genuine global transnational community with common values and a common language, now English. In the post-industrial societies of the West this community includes a far larger proportion, if not an actual majority, of their own populations than had the tiny and unrepresentative elites of previous generations. Does not this at last provide a firm foundation on which the architects of peace can now at last build a new world order?

So it does, up to a point; but two basic problems still remain.

First, the earlier supranational elites we have described may have been a tiny minority within their own societies, but they enjoyed two huge advantages. The first was the respect they enjoyed through their status as priests, as nobility, or – in illiterate societies – simply as scholars. Their social dominance was unquestioned. The second was their role – at least before the Enlightenment – in maintaining and justifying the status quo. They were not disruptive: until the eighteenth century they supported and were supported by the forces of order and tradition. But the new bourgeoisie of nineteenth-century Europe (and later in the rest of the world) enjoyed no such advantage. They had no inherited status (hence their anxiety to acquire it by assimilating with those who had) and their ideas and activities were highly subversive of the traditional order. This provoked a backlash in which the guardians of the traditional order could mobilize everyone whose way of life was being disturbed, if not destroyed, by the process of modernization and secularization those elites had set on foot. The subsequent conflict might be resolved relatively peacefully, as it was in England; it could smoulder on through generations of mutual hatred, as it did in France; it could erupt in violent civil war, as eventually it did in Spain; and it could lead, as in Germany, to actual genocide. But in Europe by the end of the twentieth century it had run its course. The opponents of modernization had, for better or worse, been defeated and the victors were peacefully co-operating across national boundaries in building their new international order. The lesson had been generally and painfully learned that societies that try to destroy such entrepreneurial talent as they possess end up on the rubbish heap of history rather sooner than their competitors.

But elsewhere the struggle continues. In many societies outside the West these bourgeois elites remain tiny minorities unrepresentative of their societies and enjoying a standard of living far beyond the reach of the mass of the peoples among whom they live. As in the Europe of the nineteenth century, their activities are regarded not as benevolent but exploitative, their values as subversive of traditional cultures; as indeed they are. They are seen as the agents of alien power, in particular one alien power, the

United States, in whose universities and those of Europe many of them have been educated. For such embattled minorities democracy is a luxury: they may need the protection and support of authoritarian regimes if they are to survive at all. This creates problems for Western liberals who simultaneously support Third-World modernization and the protection of democracy and human rights. Napoleon III, perhaps the first consciously modernizing dictator, justified his suppression of dissidents with the promise that 'Liberty would crown the edifice', an argument today used by apologists for such authoritarian regimes as China and Singapore. Often it has eventually done so, as it has in Turkey, Spain, Portugal and indeed Chile. The alternative to such authoritarianism may be not democratic institutions, but an anarchic jungle in which warlords roam free, or reactionary populist theocratic dictatorships like the Taleban. Modernization may lead ultimately to the dissemination of Enlightenment values, but it needs a framework of social and political order if it is to get going at all. Meanwhile its progress will be marked by conflicts that may have been settled in the West, but which continue to erupt everywhere else in the world: conflicts which, however localized their roots, have international repercussions.

Finally, the West continues to breed its own conflicts.

Western societies may now all be peacefully bourgeois; but bourgeois society is boring. There are certainly more sophisticated ways of expressing this, as Freud for one found, but it is a phenomenon too often overlooked by historians. As Lamartine pointed out in his explanation of *les événements* in Paris in 1848–9, it was largely boredom that destroyed the otherwise estimable monarchy of Louis-Philippe. The vast majority of mankind has never had enough leisure to experience boredom: they have had to work too hard, irrespective of age or gender, from dawn to dusk, from cradle to grave. But the medieval church knew all about it: they made *accidie* one of the seven deadly sins. Boredom with the mechanistic rationality of the Enlightenment produced the Romantic Movement and much else. Boredom has always fuelled the avant-garde in the arts. There is something about rational order that will always leave some people, especially the energetic young, deeply and perhaps rightly dissatisfied. (Anyone who taught in a Western university at the end of the 1960s will recall the scorn with which their radical students spat out the word '*bourgeois!*' as a term of abuse.) A political cause may be necessary to focus such discontent, but it does not necessarily create it. Defiance of bourgeois values may take harmless forms such as cutting dead sheep in half, or self-destructive ones, dropping out and taking to lethal drugs. But it can also lead to the demonstrative use of destructive violence or the formation of nihilistic groups. In stable societies such behaviour may only be a recurrent but manageable nuisance, but when the foundations of social stability are threatened it can become very serious; and no society can be regarded as wholly stable that has high levels of permanent unemployment or is torn by racial strife. Militant nationalist movements or conspiratorial radical ones provide excellent outlets for boredom. In combination their attraction can prove irresistible.

So although it is tempting to believe that as the international bourgeois community extends its influence a new and stable world order will gradually come into being, we would be unwise to expect anything of the kind. This was what Norman Angell and others believed in 1914: war had become so irrational a means of settling disputes that sensible people would never again fight one. But alas, they did. Let us at least hope that Kant was right, and that, whatever else may happen, 'a seed of enlightenment' will always survive.

EPILOGUE

THE REINVENTION OF WAR

On 11 September 2001 nineteen men belonging to the Islamic fundamentalist terrorist organization Al Qaeda hijacked four American airliners bound on domestic flights. Two they crashed into the twin towers of the World Trade Center in New York. A third slammed into the Pentagon building at Washington, DC, In the fourth, also bound for Washington, the passengers realized what their captors intended and forced a crash landing in Pennsylvania, killing everyone on board. Altogether some 3,000 people were massacred before the horrified eyes of a worldwide television audience. For the Americans '9–11' (as the day became known, after the all-too appropriate national telephone emergency call) was a shock greater even than that of 7 December 1941, when the Japanese attack on the US fleet in Pearl Harbor precipitated the United

States into the Second World War. There was little dissent when within a few hours President George W. Bush declared that America was 'at war'.

This book was completed more than a year before these events took place, and there is no joy in pointing out that it foresaw some of the dangers that lay ahead. It warned (pp. 95–6) that 'a loosely knit international organization of extremist *groupuscules*, inspired by the kind of religious fanaticism that had long disappeared from the modernized West but armed with the latest of modern weapons, have dedicated themselves to the overthrow of the American-led secular world order'. But this was already common knowledge. The massacre on 11 September was only an exceptionally terrible manifestation of a hostility that was widely known to exist, directed not only against the United States but against the entire culture that it symbolized: the culture of the Enlightenment, whose apparent triumph has been described in earlier pages. What did come as a surprise was that the terrorists were able to achieve their objective, not by using 'the latest of modern weapons', but with box-cutters, airline timetables and some rudimentary flight training within Western establishments. The possible use of chemical, biological and nuclear weapons remained a threat in store.

The United States declared itself to be 'at war' and reacted accordingly. The American people erupted in spontaneous outbursts of national solidarity that recalled not just Pearl Harbor but the emotional demonstrations that had swept the cities of Europe on the outbreak of the First World War. But who were they at war *with*? President George W. Bush at first defined the enemy simply as 'terror'. But terror is only a tool of strategy, not an entity that can itself be fought. The President was to broaden his definition further to include all who used terrorism as a means; not only 'non-state actors' like Al Qaeda, but governments hostile to the United States who were believed to be developing 'weapons of mass destruction', chemical, biological and nuclear, that could be used for terrorist purposes. These he specifically identified as 'an axis of evil' consisting of the 'rogue states' of Iraq, Iran, and North Korea. Members of his administration made it clear that attacks on such states should be seen as 'pre-emptive defence', and considered unilateral military action against them entirely justified. The Secretary for Defense, Donald Rumsfeld, warned the American people to prepare themselves for a period of prolonged, if not indeed perpetual, war.

These sombre predictions seemed to many to be exaggerated, but they were not unrealistic. The members of Al

Qaeda and similar organizations did indeed see themselves as being in a state of perpetual war with the United States and the cultural values that it embodied. Although their numbers were inconsiderable, they could be constantly recruited from the peoples of the Third World for whom the process of modernization was proving as traumatic as it had for Europeans a century earlier. For some of them, indeed, the promised solution to their problems – the triumph of Western prosperity and, with it, Western values – seemed a cure worse than the disease. This was not an enmity that could be resolved by the rational processes developed to deal with interstate rivalries by the disciples of Bentham and Kant. For al Queda this was not a war for rational political or economic objectives. It was a war against 'evil', that could have only an eschatological resolution. The United States understandably responded in kind.

It was not a response that evoked general approval in the West. The people of the United States may have felt themselves to be 'at war', but the rest of the world did not. America's European allies, in particular, found the American rhetoric disturbing. International terrorism of the kind revealed on 9-11 was indeed a threat that hung over them all, and Europeans already had some experience of it. Not only America's allies but also her former adversaries in

Russia and the People's Republic of China immediately offered help, with intelligence, policing, immigration and financial monitoring, as well as military assistance in eliminating Al Qaeda's principal base in Afghanistan. But over the American extension of the conflict there were many doubts. The rogue states identified by the US administration were certainly serious nuisances and their development of 'weapons of mass destruction' a legitimate cause for concern, but the American claim to a right of preemptive attack on them, and even more, to compel 'regime changes' on their governments, seemed to be stretching to breaking point the principles of international law.

As the war psychosis in the United States gradually ebbed, so the intensity of the rhetoric abated, and – under pressure not only from allies but from their own military – the administration's appetite for a military solution to the protean manifestations of terrorism diminished. The success of the campaign in Afghanistan had certainly shown that the difficulties of winning a war against Kalashnikovs with Tomahawks, described in the previous chapter of this book, could be overcome if indigenous allies could be found to provide the Kalashnikovs. But such convenient associates are not always available; while such easy victories left behind political problems from which the United States could not easily disassociate itself. Further, the support that international terrorists received from established states was less significant than their ability to establish themselves in failed or failing states against which international pressures and sanctions were of no effect. And finally – perhaps most sinister of all – the most effective supporters for Al Qaeda had come from neither rogue states like Iraq nor failed states like Afghanistan but from America's main ally in the Middle East: Saudi Arabia.

It seemed to critics of American policy, not only in Europe but increasingly in the United States itself, that its obsession with a military solution to the 'War against Terror', and in particular a pre-emptive attack on Iraq, would not only fail to tackle the roots of the problem, but might indeed make matters worse. It was generally agreed that the hatred for the West that inspired the terrorists had its roots in a malaise general throughout the peoples of the Islamic world; a malaise rooted in their inability to compete with the West by effectively modernizing their societies without adopting Western practices and values that were sharply at variance with their own. The transition from rural to urban societies was proving even more traumatic for them than it had been for Europe in the nineteenth century, given that no safety valve was any longer available in the shape of mass emigration to a New World. They had themselves to find means of absorbing into their economy a greatly swollen, newly urbanized and largely unemployed population under the age of thirty. Further, for them the whole process of modernization was not self-engendered, as it had been in the West, but an alien import, and resented as such. Islamic elites, much like those of Russia in the nineteenth century, had realized that they could solve their economic problems only through close co-operation with, if not imitation of, the West. Such elites existed even in so-called 'rogue states' – Iran, Iraq, Libya – as well as in those such as Egypt, Turkey and Pakistan that had explicitly thrown in their lot with the West and been rewarded accordingly. But even where there was no formal democracy – itself an alien concept for Islam – such governments had to take account of the bitter and endemic opposition that their policy evoked among their own peoples, not least among the immensely influential guardians of their cultural heritage, the mullahs. They had to tread a very narrow path between the co-operation with the West that alone offered economic salvation and risking overthrow by reactionary forces that regarded such co-operation as treachery.

The problem that faced the West was thus how to deal,

not simply with the threat posed by 'international terrorists' with access to weapons of mass destruction, but with the sullen hostility of an Islamic world, many of whose members, while deploring the methods used by the terrorist organizations, sympathized with their objectives and applauded the humiliation they had inflicted on the West. The destruction of the twin towers was widely seen, much as had been the fall of the Bastille in 1789, as symbolizing the humiliation of an arrogant and detested regime. That event had ushered in twenty-five years of virtually continuous warfare. It was a prospect that the American administration was now prepared to face, however daunting it might appear to its allies and associates.

Yet most European governments saw the task as not so much how to fight a war, but how to preserve a hard-won peace. Terrorism, the weapon of the weak against the strong, was an endemic problem, unlikely ever to disappear completely. Indeed, its very existence was evidence of the stability of the global system established in the wake of the Cold War. It was primarily a problem for the police and their associates, with military forces being used only in extreme emergencies. Rogue states posed specific problems that had to be dealt with in their regional context, and to intervene either to overthrow their regimes or to destroy their weapons systems would involve not only undermining the whole structure of international law that had been established since 1945, but assuming responsibilities for their governance that recent experience in the Balkans provided a strong disincentive from undertaking. Above all – and especially for Western nations that hosted large Islamic minorities – the hostility of the Islamic world was seen as something to be managed with great care. The provocation of a 'clash of cultures' of the kind foreseen by Professor Samuel P. Huntington was to be avoided as carefully as had been a nuclear exchange during the Cold War; or, if unavoidable, certainly minimized.

As was the case during the Cold War, however, the debate is taking place not only between the United States and its allies, but within the United States itself. Further terrorist atrocities on the scale of 9–11 would no doubt again silence dissenters and confirm the hard-line policy of the Bush administration, but so long as these do not occur the natural disputatiousness of that great people is likely to reassert itself; especially if the restraints on civil rights imposed in the interests of national security remain long in place. The temptation for the Americans to persist with a largely unilateral policy of an unending 'war against terror' will remain strong. In both military and economic terms, their strength is unchallengeable: insofar as a 'war against terror' can be won by military means, they can win it without help from allies. But questions arise, not only about the nature of the war, but also about the nature of the peace. After 'victory', what?

Leaving aside the question of what would constitute such a victory, there would be a natural temptation at the end of such a struggle to 'bring the boys back home' and leave the rest of the world to sort itself out, as happened after the First World War. It did not work then, and it would not work now. Like all previous victories, this one would usher in yet another new world order, and new world orders, as we have seen, need to be policed. An attractive option would be 'burden sharing': after the United States had won the war, it could leave the Europeans and other members of 'the international community' to take care of any 'nation-building' activities needed to ensure peace. But there is a limit both to the capacity and to the will of even the most co-operative of America's allies to assume any such long-term commitments. Ultimately, the United States would have once again, as after the Second World War, to assume the burden of building and maintaining peace on the foundations of the wars it had won. This could mean converting its hegemony into something more resembling an empire; taking up the Kipling-esque burden of policing the defeated territories and leading them, in spite of their protests, 'towards the light' of Western-style modernization. Like all empires, it would have to police its turbulent periphery, but unlike its predecessors it would still remain vulnerable to catastrophic blows to its centre. Such responsibilities, with all their attendant obligations, are not likely to appeal to the people of the United States. Meanwhile, regional conflicts over disputed territory in the Middle East, south Asia and Africa would continue unabated.

This is an essay about the past, not the future, and if history teaches anything, it is that historians are no better equipped than anyone else to foretell the future, and have better reason than most not to attempt it. But it seems safe to assume that if more atrocities on the scale of 9-11 recur, then the 'war against terror' is likely to be prolonged, and it is difficult to foresee when, or even how, it will end. The longer it continues, the greater the likelihood that it could escalate into a full-scale, if highly asymmetric conflict between Islam and the West. The last such confrontation lasted not a mere twenty-five years, but well over a thousand. But if terrorist activities can be kept limited by efficient international policing; if the terrorists can be isolated from the rest of the Islamic world by far-sighted economic policies and wise political management; and if rogue states can be dealt with by the international community using a mixture of sanctions, incentives and, where necessary, covert action; then the 'war' may mutate into an uneasy peace during which the root causes of terrorism can be addressed, even if they can never be entirely eliminated.

This, surely, should be the policy of the West: not to wage war, but to work, undeterred even by the most terrible setbacks, to maintain a necessarily imperfect peace; and to preserve a world order that has been gradually evolving since men first visualized its possibility, nearly three hundred years ago.